BIOCHEMICAL TERMINOLOGY

BIOCHEMICAL TERMINOLOGY

DERIVATIONS and DEFINITIONS of the UNIVERSAL LANGUAGE of BIOLOGY

Antony Mackinlay

Library of Congress Control Number:		2022918755
ISBN:	Hardcover	978-1-6698-3226-3
	Softcover	978-1-6698-3225-6
	eBook	978-1-6698-3224-9

Print information available on the last page.

Rev. date: 10/26/2022

To order additional copies of this book, contact:
Xlibris
AU TFN: 1 800 844 927 (Toll Free inside Australia)
AU Local: (02) 8310 8187 (+61 2 8310 8187 from outside Australia)
www.Xlibris.com.au
Orders@Xlibris.com.au
843617

I have never met a person who is not interested in language.
Steven Pinker in *The Language Instinct*

For if you know the origin of a word, you more quickly understand its force. Everything can be more clearly comprehended when its etymology is known.
Isidore of Seville (560–636) in *Etymologiae*

When scientists needed a new word to describe a newly discovered process or component of living organisms, they raided Greek and Latin.
John Simpson, former editor of the *Oxford English Dictionary*, in *The Word Detective*

Etymology: the origin of a particular word.
Oxford English Dictionary

CONTENT

TABLES

ACKNOWLEDGEMENTS

Thanks are due to the following people who read versions of this book and who provided suggestions and words of encouragement: Aldo Bagnara, John Carmody, Garry Graham, Wendy Glenn, Michelle Meredyth and Peter Martin.

Introduction

Names are essential. In the early part of the nineteenth century, when chemistry had shaken off the shackles of alchemy, new elements, new compounds, new techniques were discovered that needed names. It was the classical languages, Greek and Latin, to which these early scientists turned. As chemical approaches were applied to biology, much molecular terminology has also derived from these languages.

But this did not happen overnight. The synthesis of urea from ammonium cyanate by Wohler in 1828 is often cited as a turning point that demonstrated that the synthesis of molecules found in living organisms depends only on normal chemistry. But it was still possible to hold that a mysterious vital force, the *élan vital*, operated within living cells, and even Pasteur inclined to this view later in the nineteenth century. This belief could no longer be entertained after Buchner showed in 1897 that the characteristic production of ethanol and carbon dioxide during fermentation by yeast could be carried out by a cell-free extract prepared from yeast. A result of this advance, together with early work on vitamins and enzymes and ongoing chemical investigations of medical significance, was that the new discipline of biochemistry emerged. Biochemistry explains how *function* depends on *structure*: function is the bio- part, structure is the chemistry part.

The distinctive approach of biochemistry has been to follow Buchner's example and to characterize cellular components purified from cell-free extracts. This has been followed by the *in vitro* reconstruction from purified components of biological processes such as protein synthesis and DNA replication. The identification of DNA as the genetic material, the development of protein-sequencing methods and the determination of the three-dimensional structures of proteins, developments of the 1950s, meant that by then one could indeed describe biological processes in molecular terms. The success of the molecular approach is gauged by the appearance of different areas of biological research now known as molecular biology, chemical biology, cell biology, and molecular genetics as well as their numerous subdisciplines. To describe biochemistry as a universal language of biology as I do in the title of this book, is therefore no exaggeration. I am indebted to Kandel's account of his research on memory for this term.[1]

The exponential growth of biological research is reflected by the fact that barely fifty years separated the proposal for the structure of DNA by Watson and Crick and the determination of the complete DNA sequence of the human genome. As the molecular approach to the life sciences progressed what amounted to a new language was required. Languages are like organisms. They do not appear by spontaneous creation; they evolve, and so it is with the particular language of modern biochemistry which, as I will document here, largely derives from Latin and Ancient Greek root words.

Since Latin and Greek are rarely taught in schools and universities, the contribution of these languages to English and to science is not apparent. Many students studying microbiology, genetics, cell biology, botany, and zoology will encounter plenty of biochemical and molecular terminology. The same goes for the vocational life science areas of medicine, dentistry, pharmacy, physiotherapy, nursing, agriculture, and veterinary science. In many cases, lectures are often fewer than is desirable simply because there is now so much to teach and to learn. Students in these subjects should also find interest in the material here from the point of view of language but also as a study aid that will help in coming to terms with terminology.

Students who find difficulty in this respect should be inspired by the example of Francois Jacob. His studies for a medical degree in Paris were interrupted by World War II and service with the Free French Army in Africa and Europe. After recovering from serious injuries incurred after the Normandy landings, he finished his medical degree but found it hard to settle down. Eventually he was accepted for research training at the Pasteur Institute in 1950. Jacob recalled his early days at the Pasteur Institute and his problems understanding laboratory jargon as follows:

> But apart from the term *enzyme*, the words in the laboratory eluded me. I found in them nothing to grab hold of, nothing to hook on to. The words whirled around me in a sort of frenzy, but remained colourless, devoid of

meaning. They seemed to be mocking me, defying me. As
though they were guarding the portal to a temple to which
I would be refused admittance. (Jacob 1995)

In 1965 he shared with André Lwoff and Jacques Monod, colleagues at
the Pasteur, the Nobel Prize in medicine or physiology awarded for their
foundational discoveries on the regulation of protein synthesis in bacteria.

Plan of the book. Chapter 1 introduces the theme: the evolution and
structure of the English language. Chapter 2 illustrates the bare rudiments
of Latin and Ancient Greek, including tables listing the main Latin and
Greek prefixes as they have been employed in biological English. Many
readers may prefer to skip this chapter and refer back to it later. Chapter
3 enlarges on the role played by the prefixes listed in chapter 2. Chapter
4 consists of some short etymologies of technical English. In chapter 5,
biochemical nomenclature/terminology is considered under 150+ headings
and their derivation defined. This is followed by notes and references
relating to previous chapters together with the bibliography and the index.
The latter contains over 2,000 entries and, reference to the list of topics at
the beginning of chapter 5 and/or the index will enable the reader to find
the meanings and derivations of each term encountered.

CHAPTER 1

Evolution of English

The earliest language spoken in Britain that we know about belonged to the family of Celtic languages, one of many different language families spoken in Europe and as far distant as India. These, the Indo-European languages, are known to have diverged from a precursor language that was spoken somewhere in the vicinity of Turkey, 5,000–9,000 years ago, but both the time and place are controversial. Britain was a province of the Roman Empire for over three and a half centuries. Julius Caesar, during a lull in his conquest of Gaul, visited Britain in 53 BCE. This apparently was more a reconnaissance mission than a full-scale military campaign, and it was not until 43 CE, during the reign of the emperor Claudius, that the Romans came back to occupy Britain, where they stayed until 410 (Heather 2006). Not coincidentally, 410 was also the year that Alaric and the Goths sacked Rome. This was the climax of a series of incursions by the 'barbarian hordes' as the Roman Empire crumbled. Britain had already been raided by Germanic tribes from across the North Sea, and once the 'vacancy' sign went up, it was not long before the Angles, Jutes, Saxons, and Frisians came to stay and brought their various dialects with them. In time these became the language that we recognise today as Old English (OE).

By this stage, the fifth century, the original Celtic tongue had had to compete first with Latin, then with the language of the newly arrived Germanic tribes. The Celtic languages retreated to the western and northern extremities of the British Isles, where they still survive as Welsh, Irish, and Scottish Gaelic. The construction of Hadrian's Wall near the present-day border of England and Scotland shows that the Romans gave up on Scotland, but it continued to be garrisoned right up until the Romans left Britain, indicating that the Scots never gave up on the Romans. So while we can understand what happened to Celtic, it is surprising that Latin left no significant mark on what became OE. The Romans had a long-standing policy of imposing their language as well as the yoke on their conquered peoples; witness how Latin evolved into the Romance languages in Italy, Spain, France, and elsewhere. The difference in Britain must have been that the Germanic tribes, who were now in charge, had never been colonized by the Romans. As far as they were concerned, Latin was a foreign language for which they had no use and which they proceeded to erase from national memory. Only a handful of Latin words survived the turmoil following the departure of the Romans. OE very nearly suffered the same fate following the Norman conquest in 1066.

During the 500 years or so before the Normans arrived, OE consolidated and began to change from an inflected language, where word endings indicate subject and object, to one where word order serves this purpose. A further development during this period was the reintroduction of Latin with the arrival of Augustine and his Irish missionaries in 597. Within a year, they had converted King Ethelbert of Kent, which facilitated the spread of Christianity. While mainly the clergy and the monasteries used Latin, the crucial contribution to English that followed on from their activities was the spread of literacy. Thus, the Venerable Bede (672–735), a Northumbrian monk, produced a history of the English Church as well as other writings. Bede wrote in Latin at a time when across the Channel in Europe, few people were able to speak or write classical Latin, and literacy in general was rare. Anglo-Irish scholarship was one of the few lights during this so-called dark age in Europe, which extended from around 550 to the crowning of Charlemagne in 800 as head of the Holy Roman Empire.

A major challenge both to English identity and to the language came with repeated incursions by Danish and, to a lesser extent, Norwegian raiding parties that culminated in battles involving thousands of soldiers on both sides in 878. The result was a stalemate. The Danes controlled the northern half of the country while the English king Alfred, with his capital at Winchester, controlled the south. The dialects spoken by the Viking invaders, together with the words that infiltrated English and which still survive, are known as Old Norse (ON). Many place names such as those ending in *-by*, *-thorpe*, and *-thwaite* and family names which end in *-son*, such as Harrison, Robson, and Johnson, reflect the power and authority the Vikings exerted until the Norman invasion. Nevertheless, relatively few ON words made it into standard English. Bragg summarises the composition of the language prior to 1066 as follows: it contained a score of Celtic words, 200 Roman words, and 150 ON and the rest Germanic in a total lexicon of about 25,000.[2] A reasonable conclusion would be that by then what we now refer to as OE was sufficiently comprehensive and robust that it resisted being overpowered by a foreign tongue. It did then survive the onslaught of Norman French, but only just, and it took 300 years for the dust to settle. What emerged was a much-altered language.

1066 and ALL THAT

The Normans who invaded in 1066 and defeated the English at the Battle of Hastings were Scandinavians who two centuries previously had settled in what became known as Normandy. They were Old French (OF)-speaking cousins of the Vikings who had dominated the north of Britain. Although they ruthlessly asserted their authority and expropriated the land, the Normans did not bother to suppress the language of the people whom they ruled, but it was French they used to administer Britain. Bragg[3] lists the areas where French words prevailed and have become part of modern English: *art, architecture and building, Church and religion, entertainment, fashion, food and drink, government and administration, home life, law and legal affairs, scholarship and learning, literature, medicine, military matters, riding and hunting and social ranking*. Such words of OF derivation had in turn evolved from a vernacular version of Latin. This massive infusion of OF words enriched English immeasurably.

The flip side of this was that during the following 300 years, until everyone again spoke some form of English, about 85 per cent of the OE vocabulary was lost. Assuming there were about 25,000 OE words before 1066, barely 4,000 OE words survived. Those survivors are mainly words that describe everyday objects: body parts such as *arm, head, gut, kidney, liver*; animals such as *cow, horse, sheep*; and aspects of the countryside such as *field, hill, tree, wood*. Another relatively small group of residual OE words are the ones that define English grammar. English speakers still rely on these to communicate with each other. The 100 most commonly used words include verbs such as *be, can, do, get*; pronouns such as *who, which, him, them*; prepositions such as *in, of, to, with*; the indefinite article *a* and the definite article *the*. These were the words that the common people used, while the people in authority spoke French. English having eventually won out, these are still the basic tools that modern-day English speakers use. Table 1 lists the 100 most commonly used English words.[4] Leaving aside the three ON words in that list, there are only two, *oil* and *number*, that have come into English from Latin roots via OF. No such list is definitive. Another list compiled by Oxford Online, based on all written material in English, has the same first 50 words in somewhat different order to that in table 1 except for *say* instead of *said* and *me* as well as *my*. It is claimed that these top 100 account for about one half of all written material. A characteristic of OE words is that they are short words.

A process that had operated on the various versions of English over the course of several hundred years was one which eliminated most of the variation in word endings (called inflections), which is a feature of the classical languages. In an inflected language, word order is not important because the subject and object are distinguished by the word endings of the respective nouns, referred to as noun declensions. In English, without this variety of word endings, the word order is important: subject, verb, object. In English, noun plurals are indicated by an *s* suffix: one *house*, two *houses*. Barely a handful of irregular plurals have survived: *mouse* and *mice*, *ox* and *oxen*, and one *sheep* and two *sheep*.

For an inflected language, verb endings (conjugation) indicate when the action is happening: now, in the future, or in the past. English does this very precisely by using auxiliary verbs: I call, I am calling, I *will* call, I *did* call, I *was* calling, I *will have* called, I *will have been* called (or calling), I *should have been* called, etc. The past tense of nearly all verbs is indicated by the suffix -*ed* as in I *call* and I *called*. English still has about 180 verbs with irregular forms of the past tense. These are survivors from OE, such as *buy* and *bought*, *send* and *sent*, *do* and *did*, *drink* and *drank*. An important consequence of the nearly complete absence of noun declension and verb conjugation was the invention of a long list of suffixes that must have accounted for a significant expansion of the English lexicon. A selection of English suffixes is in table 2.

English is a simple language compared to an inflected language, and another great advantage which it enjoys, compared to French and German, is that it has dispensed with noun gender. In French for example, *glucose* is feminine but *glycogen*, a polymer of glucose, is masculine, while in modern German, *glucose* is feminine, *glycogen* is neuter, and *girl* is also neuter!

English could have disappeared after the Norman Conquest, to be replaced by French. However, this did not happen. The Normans and the native English intermarried and raised bilingual children. Another factor was that in 1218, the now-English Normans lost to the French Crown the lands that they had previously held in Normandy. Norman old-style French (OF) as spoken in England diverged from the French spoken in France. The descendants of the original invading Normans thus came to identify increasingly with the previously Anglo-Saxon locals, especially once the two ethnic and language groups had intermingled for many generations.

In 1348, the plague, the Black Death, caused by the bacterium *Yersinia pestis*, broke out in England. Up to a third of the population, possibly 1.4 million people, succumbed. The resulting shortage of labour meant that the surviving English-speaking lower classes gained more influence. It was in 1362 that English was finally recognized as the language of official business. In 1399 when King Richard II abdicated, Henry IV was crowned, with the ceremonies conducted in English. Between 1382 and 1395, copies appeared of the Bible translated into English, the Wycliffe Bible. Meanwhile, Chaucer, who was born in the 1340s and died in 1400, had written the Canterbury Tales, which with a little difficulty can be read and understood by a modern English reader. The period from 450 to 1150 is regarded as OE, from 1150 to about 1500 as Middle English (ME), while the period subsequent to 1500 is Modern English.

LINGUA FRANCA: THE OLD ONE and THE NEW ONE

A *lingua franca* is defined as a language used for communication between two or more groups that have different native languages. English has rather suddenly become the global lingua franca. The term arose among traders in the Mediterranean in medieval times, *franca* being a Latinized version of words used by Greek and Arabic speakers, for whom all Western Europeans were Franks. In Europe for nearly a thousand years, the lingua franca was Latin. Usually, Latin is described as a dead language, but even today Latin shows up in unexpected places: the American actress Angelina Jolie has had the Latin phrase *quod me nutrit me destruit* (what nourishes me also destroys me), tattooed provocatively on her lower abdomen. This apparently has started a trend, with results not always up to the grammatical standards of Cicero. The point about Latin is that while it is a long time since anyone spoke it as a first language, its essential structure has remained fixed. The main reason that Latin was widely used and understood in Europe by about the year 1000 was that what had started out as various Latin vernaculars had by that time developed into distinct, mature languages: French, Italian, Spanish, and others, i.e. the Romance languages. Latin was how speakers of different Romance languages, along with people speaking English and Germanic languages, communicated with each other.[5]

In retrospect, it seems almost inevitable that with the advent of mass travel and rapid global communication, some global language or lingua franca would emerge in order to allow communication in a multilanguage world, a problem that was noted in the Bible in connection with the Tower of Babel. Other examples include Greek, which was the lingua franca for a thousand years in the Eastern Mediterranean, Arabic following the spread of Islam, and Chinese in East Asia. The use of Latin as a European lingua franca declined with the rise of the nation-state. The last two centuries can be regarded as an interlude between the Decline of one lingua franca and the appearance of another. A lingua franca has traditionally been the language of scholarship, as Latin once was, and this is also the case with English.[6]

THE MEDIAEVAL REVIVAL of CLASSICAL LATIN

During the three hundred years following the end of the Western Roman Empire, most of its former territories, corresponding to modern Italy, France, and Spain, were divided up among kingdoms whose rulers spoke Germanic dialects. The result was that while vernacular Latin continued to be spoken in these countries and eventually evolved into the Romance languages, there was a general decline in literacy. Scholarship based on written classical Latin, apart from a small number of isolated scholars, was confined to the church in Rome as well as in England and Ireland. Things only began to change in mainland Europe with the arrival of the Englishman St Boniface (675–754), known as the Apostle of the Germans. Boniface reformed the Church in the kingdom of the Franks, ruled at that time by Charles Martel (686–741). It was Martel's grandson, Charlemagne (747–814) who reunited Western Europe within the Carolingian empire.[7]

Charlemagne soon set about reorganizing education and called on the intellectual resources of the Church to help achieve this aim. It was this initiative that led to Latin becoming the lingua franca of Europe for the best part of a thousand years, the Latin millennium. Charlemagne saw himself as the successor to the Roman emperors and was crowned emperor of the Holy Roman Empire by the pope in 800. Interestingly, the last of the Holy Roman emperors abdicated in 1806, about the same time as Latin lost its status as the language of communication. The suggestion has been made that the twenty-odd members of the EU could do worse than conduct their meetings in Latin.

THE CLASSICS REVISITED

The other development that led to interest in the classical languages was the gradual rediscovery of ancient writings, both Greek and Latin, that had been lost for hundreds of years. The libraries of monasteries throughout Europe were scoured for these classical works, and naturally, it was the clergy who initially dominated the scholarship that ensued. During the next several centuries, Europe as a whole became more prosperous, new land was opened up for food production and the population increased. This new prosperity resulted in an extraordinary outpouring of religious zeal leading to the construction between 1050 and 1350 in France alone of some eighty cathedrals and thousands of smaller churches. This was accompanied by a comparable intellectual ferment, often located within cathedral schools and later in universities. At the same time as this was happening, several Crusades were undertaken in unsuccessful attempts to recapture Jerusalem from the Muslims.

It is sometimes claimed that this period saw the beginnings of science. There was certainly plenty of philosophical speculation. What did distinguish the thirteenth and fourteenth centuries from what had gone before was a willingness to think about the limitations of faith and reason and an interest in the natural world. The term *natural philosophy*, which was a prerequisite for science to develop, arose from this, but the crucial step of experimentation was not taken. It was a case of developments that were necessary but not sufficient. Names that are mentioned in this context are Robert Grosseteste (1175–1253), Roger Bacon (1214–1294), and William of Ockham (1287–1347). A multidisciplinary investigation of the writings of Robert Grosseteste has recently been carried out at the University of Durham.[8] Like his contemporaries, he was interested in the nature of light and vision, thought that something like gravity must exist and that the world began with a large explosion. Francis Bacon (1561–1626), a contemporary of Shakespeare, who insisted on experimentation, is credited with being the first true European scientist.

ENGLISH GETS LATINIZED

English seems to have a natural appetite for assimilating words from other languages. This became apparent as British traders and the British Empire extended their reach. Ever since the Norman invasion, words derived from Latin had been adopted into English, but in the sixteenth century what had been a hobby became an obsession. A vast number of Latin words were anglicized to the point where today it would be a challenge to write a few lines without including words that were newly minted at that time. Many related Latin-derived words, now with altered meanings and spellings mangled by the French, had already arrived in English with the Normans. As well, words had come in from church Latin off and on over the centuries ever since the missionary Augustine arrived in 597. By the fifteenth century, the printing press had been invented, and printing spread through Europe like—well, like personal computing in the late twentieth century. The result was an avalanche of books in English that included most of the ancient Latin texts, many available in English for the first time. This served to accelerate the generation of English words from Latin roots. At the time, not everyone was happy about this influx of Latin-based words, but unlike the 1066 event, this incursion by Latin simply enriched the English language without seriously threatening it. This was the last major infusion into the English language, but it has continued to evolve at a rapid rate ever since. The addition of classically derived words to English helped to fill a need stemming from the increasing sophistication and diversity of society and, in the long run, probably amplified these qualities.

Ancient Greek words never infiltrated English in the way that Latin did. Greek texts were originally studied in Latin translations, and it appears that the first translation of Homer's *Iliad* directly into English was not published until 1611, followed by the Odyssey five years later.[9] Before then, Britain was essentially 'Greekless', apart from an indirect source of Greek, the sizeable number of Greek words, including many medical terms, that came into English via Latin. As will become apparent here, words derived from both Latin and Greek find extensive use in the language of modern science. Table 3 illustrates how both Greek and Latin roots have become part of modern English used to describe aspects of human anatomy.

Table 1. The 100 most frequently used English words
Numbers indicate position in the top 100. ON: derived from Old Norse.

a (4)	get (94)	my (81)	there (41)
about (54)	go (74)	no (77)	these (59)
all (34)	had (28)	not (32)	they (19) (ON)
an (43)	has (69)	now (90)	this (23)
and (3)	have (24)	number (76)	time (68)
are (15)	he (11)	of (2)	to (5)
as (16)	her (62)	oil (88)	two (71)
at (21)	him (65)	on (14)	up (52)
be (22)	his (18)	one (27)	use (42)
been (85)	how (48)	or (26)	was (12)
but (31)	I (20)	other (53)	water (84)
by (29)	if (50)	out (55)	way (78)
call (86)	in (6)	part (100)	we (36)
can (39)	into (67)	people (80)	were (35)
come (97)	is (7)	said (40)	what (33)
could (79)	it (10)	see (75)	when (37)
day (95)	its (89)	she (46)	which (45)
did (47)	like (66)	so (60)	who (87)
do (47)	look (70)	some (61)	will (51)
down (93)	long (92)	than (82)	with (17)
each (44)	made (98)	that (9)	word (30)
find (91)	make (64)	the (1)	would (63)
first (83)	many (56)	their (49) (ON)	write (73)
for (13)	may (99)	them (58) (ON)	you (8)
from (25)	more (72)	then (57)	your (38)

Table 2: English suffixes

-able	—	adjectives: available, controllable, doable, predictable, solvable, suitable
-age	—	nouns: appendage, lineage, mileage, percentage, spillage, tonnage
-al	—	adjectives: ancestral, biological, dorsal, hormonal, logical, microbial, vestigial
-an	—	complex carbohydrates: dermatan, dextran, glycosaminoglycan, keratan, mannan
-ance	—	nouns: appearance, assistance, dominance, importance, inheritance, resistance
-ane	—	usually indicates a saturated hydrocarbon: methane propane, cyclohexane
-ar	—	mainly for forming adjectives: globular, linear, regular, similar, but also: scholar
-ary	—	adjectives and nouns: elementary, evolutionary, tertiary, also ovary, salary
-ase	—	suffix indicating an enzyme but excluding some early ones like pepsin, trypsin
-ate	—	salts: acetate, phosphate, sulphate, etc., also: dialysate, eluate, filtrate, precipitate
-ate	—	verbs derived from 1st conjugation Latin verbs: activate, regulate, differentiate
-ation	—	nouns derived from 1st conjugation verbs: activation, regulation, differentiation
-en	—	forming verbs from adjectives: deepen, lengthen, moisten, widen
-ence	—	nouns: difference, excellence, resilience, science, transience, virulence
-ene	—	typically indicating unsaturated hydrocarbons: acetylene, benzene, ethylene
-gen	—	indicating precursor or cause: carcinogen, collagen, halogen, hydrogen, zymogen
-genesis	—	describing a process: carcinogenesis, gluconeogenesis, mutagenesis

-genic — pertaining to or producing: glucogenic, ketogenic, hallucinogenic, photogenic

-ic — forming adjectives: aquatic, aerobic, genetic, also: ferric, sulphuric, hydrochloric

-ical — forming adjectives: biological, chemical, hypothetical, pathological

-ide — various chemical names: amide, chloride, cyanide, glycoside, nucleotide, peptide

-ify — forming verbs: acidify, amplify, emulsify, purify, saponify, verify

-ile/-il — adjectives: contractile, fertile, juvenile, tensile, also nouns: fibril, tendril

-lity — forming nouns from adjectives: contractility, fertility, permeability, solubility

-in — names of chemicals/proteins: actin, myosin, penicillin, protein, rennin, trypsin

-ine — all amino acids except tryptophan, nucleic acid bases adenine, thymine, etc.

-ist — mainly occupations: botanist, chemist, zoologist, etc.

-itis — mainly diseases: arthritis, nephritis, tonsillitis, etc.

-ity — forming nouns: activity, affinity, homogeneity, obesity, polarity, specificity

-ium — mainly names of metals: sodium, potassium, etc., also: helium, ammonium

-ive — forming adjectives: active, inventive, native, negative, also: sedative

-ise/-ize — forming verbs, spelling variable: internalize, synthesize, theorize, visualize

-ment — nouns often indicating action: attachment, measurement, movement

-oid — adjectives and nouns: amyloid, anthropoid, colloid, diploid, opioid, steroid

-ol — usually an alcohol or reduced derivative: butanol, glycerol, mannitol, phenol

-ose	—	usually a carbohydrate: glucose, cellulose, etc., also: adipose, comatose, verbose
-osis	—	a process or condition: meiosis, metamorphosis, mitosis, neurosis
-ous	—	indicating lower valency: ferrous, nitrous, sulphurous, also: fibrous, gelatinous
-sity	—	forming nouns: curiosity, generosity, heterozygosity, verbosity, viscosity

Table 3. List of OE names for body parts with Latin and Greek derivatives

	Latin	Greek	Examples of derived English words
Windpipe (OE)	trachia	bronkhos	trachea, bronchitis
Heart (OE)	cor, cord-	kardia	coronary, cardiac
Lung (OE)	pulmo	pneumon	pulmonary, pneumonia
Jaw (ME from OF)	maxilla	gnathos	maxillary, gnathostomes
Bladder (OE)	vesica	kystis	vesicle, cystitis
Arm (OE)	bracchium	brachion	embrace, brachiopod
Mouth (OE)	os, or-	stoma	oral, stomate
Hand (OE)	manus	cheir	manipulate, chiral
Rib (OE)	costa	pleuron	intercostal muscle, pleurisy
Skin (OE)	cutis	derma	subcutaneous, dermatitis
Tongue (OE)	lingua	glotta	bilingual, epiglottis, polyglot
Breast (OE)	pectus, pector-	stethos	pectoral muscle, stethoscope
Vein (ME from OF)	vena	phleps, phleb-	venous, phlebitis
Mind (OE)	mens	phren, psyche	mental, schizophrenia, psyche
Blood (OE)	sanguis	haima	sanguine, haemoglobin
Belly (OE)	venter	gaster	ventral, gastric
Back (OE)	dorsum	noton	dorsal, notochord
Head (OE)	caput	kephale	precipitate, encephalitis
Finger (OE)	digitus	daktylos	digital, perrissodactyl
Kidney (?)	ren, renes	nephros	renal, nephritis
Foot (OE)	pes, ped-	pus, pod-	bipedal, tetrapod
Eye (OE)	oculus	ophthalmos	binocular, ophthalmology
Nose (OE)	nasus	rhis, rhin-	nasal, rhinoceros

CHAPTER 2

The Classics

As already explained, the broad outline of English was heavily influenced by Latin. English was never Hellenized in the way that it was Latinized. The bulk of English words derived from Greek are scientific and technical terms that have been introduced relatively recently. There are also words that were adopted from Greek into Latin and then into English. Many of these are medical terms, because Rome appears to have relied on Greek physicians who, like Galen, wrote extensively in Greek. You can get an idea of the influence of Greek on Latin at this time by looking up words in a comprehensive Latin dictionary that begin with the Greek prefixes listed in table 6. Over the years, both languages have been mined for terminology required by the developing sciences. Here you will find the rudiments of ancient Greek and Latin and how prefixes were used to expand those languages.

The Latin alphabet, like the Greek, has twenty-four letters and is the same as English except it is missing *j* and *w*. Listed in table 4 below are the twenty-four letters of the Greek alphabet with some comments on its transliteration into English. Latin and Greek, with their alphabets obviously related, can be described as sister languages, both derived from the original proto-Indo-European. Students of the biological literature soon become familiar with numerous Greek letters. Different DNA polymerases are designated α-, β-, γ-, etc.; the three families of opioid receptors are the μ-, κ-, and δ-receptors.

Conversion of an ancient Greek word to English, or transliteration, follows that given in the 'English equivalent' column in table 4. Thus, μελανχολια becomes **melancholia**, or in English, the more familiar *melancholy*. Greek has no *h* but a diacritical mark, also called a breathing, on an initial vowel indicates it is to be aspirated. Greek for blood αἷμα transliterates to **haima** and to *haem* in English or *heme* in American English. Diacritical marks in general indicate differences in pronunciation; **diakritikos**, the adjective, comes from **diakrinein** 'to distinguish' while the verb **krinein** 'to separate' also provides the root for *endo<u>crin</u>ology*—see later under the entry for

SECRETION MECHANISMS. The -**ikos** suffix of **diakritikos** is a typical adjectival ending; -**ein** as in **krinein** often denotes the infinitive form of a verb. The absence of *h* in the Greek alphabet seems at first a little odd since the age of classical Greek culture is referred to as the Hellenistic age. English also has a problem with *h*—sometimes it is pronounced and sometimes not, as in 'I spent an hour at a hospital.'

Table 4. The Greek alphabet, after Morwood (2001)

Greek letter	small	capital	English equivalent
alpha	α	A	a as in awake
beta	β	B	b as for English b
gamma	γ	Γ	g as in go
delta	δ	Δ	d as for French d, tongue on teeth
epsilon	ε	E	e as in pet
eta	η	H	e as in air
theta	θ	Θ	th as in thin
zeta	ζ	Z	sd as in wisdom
iota	ι	I	i short as in fit, long as in read
kappa	κ	K	c hard as in cat
lambda	λ	Λ	l as in leap
mu	μ	M	m as in met
nu	ν	N	n as in net
xi	ξ	Ξ	x as in box
omicron	o	O	o short as in pot
pi	π	Π	p as in pot
rho	ρ	P	r rolled r
sigma	ς, σ	Σ	s as in sing
tau	τ	T	t as in stop
upsilon	υ	Y	u or y as in loot or synthesis
phi	φ	Φ	ph as in foot
khi	χ	X	ch as in loch
psi	ψ	Ψ	ps as in lapse
omega	ω	Ω	o as in sow

For sigma, it is ς at the end of a word and σ at other positions.

One thing about the Greek alphabet: it explains the origin of the word *alphabet*. The Greek alphabet was derived from the earlier Phoenician alphabet. Local variations became standardized towards the end of the fourth century BCE, although the lower-case letters shown here did not come into use until much later. This is still the alphabet used in modern Greece.

It is said that Greek verbs can have up to 300 different forms (Pinker 1994, Andrews 1947). Don't panic. Here we are concerned only with the roots or stems of those words that have come down to English. For Greek verbs in most cases it suffices to quote the infinitive. This is because when scientists derive new English words based on Greek verbs, it is either the infinitive or the first person present tense, or failing that, the corresponding noun that they are thinking of. Thus, **skopein** 'to look at' provides an obvious root for *telescope* and *microscope*; the **tom** root in *entomology* regarded as derived from **temnein** 'to cut' is not quite so obvious, but **tomos,** the noun meaning 'cut', is obvious. See the entry for **METABOLISM** for its derivation.

An example of a Greek noun with a number of English progeny is **ergon,** meaning 'work'. In science, the *erg* is a unit of energy equal to 10^{-7} joules, the latter being the SI (Système Internationale) unit. A corresponding adjective is **energos**, with The **en**, prefix meaning 'busy', from which English *energy* derives, also *energetic* with the suffix corresponding to Greek **-ikos** added. A manual worker was **ergocheiron**, in which the word for 'work' combines with **cheir,** the word for 'hand'. Stepping up the social scale, we come to **cheirourgos,** someone else who works with their hands, and this is the origin of the English word *surgeon*. Other words based on the **ergon** root that you will come across in these pages or elsewhere include *exergonic, endergonic, cholinergic, adrenergic, bioenergetics, synergy*, and *argon*. **Cheirourgos** is an example of a compound noun, which the Greeks were fond of inventing, but which does not occur in Latin. More familiar compound nouns are **hippopotamus** or 'river horse' from **hippos** 'horse' + **potamos** 'river' and **rhinoceros** or 'nose horn' from **rhis, rhin-** 'nose'+ **keras** 'horn'.

ASPECTS of LATIN

A large number of English words derive ultimately from Latin verbs. Most Latin verbs belong to one of four groups or conjugations. These are first conjugation: infinitives ending in -are: e.g. **vocare**: to call second conjugation: infinitives ending in -ēre: e.g. **movēre**: to move third conjugation: infinitives ending in -ere: e.g. **solvere**: to loosen fourth conjugation: infinitives ending in -ire: e.g. **audire**: to hear.

The ē in the second conjugation indicates the e is pronounced long as in *era*, but we are not concerned with pronunciation here.

When it comes to words derived from a verb, the derived word can be said to be based on a particular stem. Verbs such as those above have three separate stems: a present stem based on the present infinitive 'to call' and a perfect stem from the first-person perfect tense 'I called' as well as the passive perfect participle (ppp.) 'having been called'. The significance of these stems is mainly that they provide the basis for construction in Latin of other tenses such as the future and subjunctive. The table below illustrates how English words have arisen based upon different word forms of the Latin. For the verbs listed above, the stems are indicated here in bold:

	To call etc.	I called etc.	Having been called
Voco—I call	**voca**-re	**vocav**i	**vocat**-um
Moveo—I move	**movē**-re	**mov**i	**mot**-um
Solvo—I loosen	**solve**-re	**solv**i	**solut**-um
Audio—I hear	**audi**-re	**audiv**i	**audit**-um

From the infinitive stems, we get *vocal, move, solve, solvent, audible,* and *audio.* From the passive perfect participle column: *advocate, vocation, motion, motor, solution, audition.* When it comes to discussing the derivation of English words from both Greek and Latin, rather than specifying a particular stem, it is perfectly adequate in most cases to stick with the infinitive and say that the words *vocal* and *vocation* derive from the Latin root **vocare.**

Some Latin verbs and their English offspring:

fero, ferre, tuli, latum: an obviously irregular verb with the meaning 'to bear, carry or bring', the descendants of which include *confer, defer, infer, prefer, refer, transfer, circumference, afferent, efferent, suffer, ablate, collate, elate, oblate, prolate, relate*, and *translate*. Latin **ferre** is related to the Greek **pherein** 'to bear', from which many English words, such as *electrophoresis, phosphorus, pharmacophore*, and *chromophore* derive.

pono, ponere, posui, positum: 'to put or place': *apposite, component, compose, composite, composition, compound* (OF), *depose, dispose, disposition, expose, expound* (OF), *exponent, impose, indisposed, interpose, oppose, position, positive, postpone, preposition, proponent, proposition, propound, repose, superimpose, suppose, transpose*

verto, vertere, verti, versum: 'to turn': *adverse, averse, advert, advertise, controversy, converse, convert, divert, diverse, evert, extrovert, introvert, inverse, invert, perverse, pervert, reverse, revert, subvert, transverse, verse, version, vertebra, vertebrate, vertical, vertigo, universe, university*

volvo, volvere, volvi, volutum: 'to roll or turn around': *convoluted, devolve, evolution, evolve, involve, revolution, revolve, voluble, volume*

Knowledge of words derived from the same root helps to expand your scientific vocabulary.

PREFIXES and SUFFIXES

Both Romans and Greeks used many prefixes to expand their vocabulary, and this has carried over into English. Table 5 consists of a nearly complete list of Latin prefixes together with some English words that contain these prefixes. It is safe to assume that any English word that contains one of these prefixes derives from a Latin root, and a similar rule, with only a few exceptions, applies to words with Greek prefixes. Table 6 lists some Greek prefixes. Most of the words in these tables are words that one might meet in a scientific textbook or paper.

In most cases, the addition of a prefix to a Latin stem was already a feature of Latin and usually the prefix + root came across into English together. Often a stem joined to multiple prefixes has generated a word family in both languages. An example is the family of related verbs based on **capio, capere, cepi, captum** meaning 'seize' or 'capture' as well as a variety of other meanings depending on context. From **capere** we get **recipere, recepi, receptum**, meaning 'to take back' and from which we get the English words *receptor, reception,* and via OF, *receive* and *receipt.* In similar fashion from **concipere:** *concept, conception, conceive, conceit;* from **decipere:** *deception, deceive;* from **excipere:** *except, exception;* from **percipere:** *percept, perception, perceive;* from **intercipere:** *intercept, interception.* English achieves a further round of language amplification by adding additional prefixes and suffixes to obtain words such as *imperceptible, inconceivable, unexceptional,* etc. See examples of English suffixes in table 2 in chapter 1. English is riddled with word families in which a stem is joined to a variety of prefixes and suffixes.

The verb *to replicate,* formed from **re + plicare,** comes with biological credentials and belongs to a similar extended family, the root meaning 'to coil' or 'to fold'. English also has *complicate, duplicate, explicate, implicate, supplicate,* and via OF, *multiply* (multiplication) and *apply* (application). These last two examples illustrate a relatively common situation in which the verbs came via OF while the corresponding nouns were imported directly from Latin. Table 7 is a list of Latin words that have come across directly into English usage, most of which have retained their original meaning.

You will notice in table 5 where Latin prefixes are listed that for the case of *ad–* it forms words such as *admit, adsorb, advise* in which the prefix joins directly to one or other root words. But in that list there are also words such as *accept, affect, agglutinate, alleviate, apparatus,* and *assume.* These are examples of what grammarians refer to as *assimilation,* in which the *d* of *ad* is replaced by the first letter of the root word. The reason this has come about is that it makes the word easier to say and to listen to as well. Other examples occur with the prefixes *com–, ex–, in–, ob–,* and *sub–.* The word *assimilation* is also an example of assimilation.

The main dictionary references used here for the classics are more recent versions based on Lewis and Short (1879) and Liddell and Scott (1889).

Table 5. A selection of words containing Latin prefixes

ab-	abbreviate, aberrant, ablate, abnormal, abrupt, abscess, absent, absolute, absorb, abstract, abundant
ad-	accept, accumulate, accuracy, addict, adduct, adjunct, adjuvant, admit, adsorb, adverse, advise, affect, afferent, affinity, agglutinate, aggregate, alleviate, allocate, apparatus, apposite, approximate, ascribe, associate, assume, attenuate, attract
com-	collaborate, collate, collect, combine, communicate, compare, competent, complement, complementary, complex, component, compound, comprehend, compute, concatenate, concentrate, concept, condense, confer, conformation, congenital, conjugate, consistent, construct, contagious, convalesce, convection, co-opt, coordinate, correct, correlate, correspond, corroborate
contra-	contraceptive, contradict, contraindicate, contrast, contravene, controversy
de-	decay, deceive, decide, decrease, deduce, defect, defer, deficiency, deform, deflect, defuse, degrade, deliquescent, demotion, denature, depend, depress, derive, describe, descend, deserve, desist, destroy, detect, detergent, detoxify, devour
dis-	disaggregate, discern, discharge, discontinue, discount, discourse, discover, discuss, disinfect, disintegrate, dismantle, disorder, dispense, disperse, dispose, dispute, disrupt, dissect, dissociate, dissolve, distant, distinguish, distribute, disturb
ex-	effect, efferent, efficient, effluent, eject, elucidate, elude, elute, emit, event, eviscerate, evolve, exact, exceed, except, excise, exclude, excrete, exit, expect, expel, expire, explicit, expose, exponent, expound, express, extinct, extract
extra-	extracellular, extrachromosomal, extracurricular, extraordinary, extrapolate.
in-	means 'not' when present as a prefix for a very large number of words, such as *inactive*, *inexact*; it can also mean 'in or into' as in *inject*. See 'Prefixes' chapter.
inter-	interact, intercalate, intercellular, interest, interface, interior, intermediate, intermolecular, interpolate, intersect, interval

intra-	intracellular, intramolecular, intramuscular, intraspecific, intravenous
multi-	multicellular, multidisciplinary, multifactorial, multiple, multivalent, multiplication
ob-	object, observe, obstacle, obtain, obvious, occupy, occur, offend, opportune, oppose
per-	perceive, perennial, perfect, perfuse, permanganate, permissive, permutation
pre-	precancerous, precaution, precedent, precipitate, precise, preclinical, preconceived, precursor, predict, prefer, prefix, preliminary, prepare, preproenzyme, prescribe, preserve, presuppose, presynaptic, previous
pro-	procedure, product, proficient, profile, progeny, progress, proinsulin, project, prolactin, prolate, promoter, pronucleus, propagate, proportion, proposal, provenance
re-	react, reagent, recede, receive, receptor, recruit, recur, reduce, refer, relate, reflect, reflex, reflux, refraction, reject, relevant, repair, replete, replicate, repress, resist, resolve, respect, respire, resume, retain, retract, revive, reverse, revise, revolve
retro-	retrograde, retrovirus, retrospect, retrotransposon
se-	secede, seclude, secrete, secure, seduce, separate, segregate
sub-	subconscious, subcutaneous, subdivide, subject, sublime, subliminal, subscript, subsequent, substance, substitute, substrate, subtract, subunit, succeed, succinct, sufficient, suffix, suggest, suppose, suppress, surrender
super-	supercoil, superficial, superfluous, superimpose, superinfection, supernatant
trans-	transcription, transduction, transfection, transferase, transformation, transgenics, transition, translation, translocation, transposons, transversions

Table 6. A selection of words containing Greek prefixes

acro-	**akron** 'top', **akros** 'at the end, tip, top': acrocentric, acronym, acrobat, acrostic
allo-	**allos** 'other, different': allotrope, allograft, allopurinol, allosteric
amphi-	**amphi** 'around, both, double': amphibian, amphitheatre, amphoteric, amphipathic
ana-	**ana** 'up, back, again': analysis, analogy, anatomy, anabolic, anachronism
anti-	**anti** 'against, opposed to': antibiotic, antibody, anticodon, anticoagulant, antidote
apo-	**apo** 'from, away from': apoptosis, apoplexy, apoprotein, apogee, apocrine
auto-	**autos** 'self': autosomal, autonomous, autoimmune, autologous, autolysis, autopsy
bio-	**bios** 'life': bioassay, biochemistry, biodegradable, biology, biopsy, biosynthesis
cata-	**kata** 'down, against, very': catabolism, catalysis, catalepsy, cataract, catarrh
dia-	**dia** 'through, between, across': diabetes, diagnosis, diagonal, diagram, dialysis
dys-	**dys** 'bad, difficult': dysentery, dysfunction, dyslexia, dyspepsia, dysplasia, dystrophy
ecto-	**ekto** 'outer, external': ectoderm, -ectomy, ectoparasite, ectopic
en-	**en** 'in, within': enantiomer, encephalitis, endemic, energy, entropy, enthalpy, enzyme
endo-	**endon** 'within': endocardium, endocrine, endocytosis, endoderm, endogenous
epi-	**epi** 'upon, in addition': epidemic, epidermis, epigenetic, epimer, epinephrine, episode
eu-	**eu**: 'good, well, normal': eubacteria, eukaryote, euchromatin, eulogy, euploid

hetero-	**heteros** 'other': heterochromatin, heterocyclic, heterogeneous, heterozygote
holo-	**holos** 'whole': holocaust, Holocene, holoenzyme
homo-	**homos** 'same': homocysteine, homogenate, homogeneous, homologous, homozygote
hyper-	**huper** 'over, above, beyond': hyperglycaemic, hyperimmune, hyperplasia, hypertonic
hypo-	**hupo** 'under': hypochondria, hypodermic, hypoglycaemic, hypothalamus, hypothesis
idio-	**idios** 'own, distinct': idiom, idiopathic, idiosyncrasy, idiotype
iso-	**isos** 'equal': isoelectric, isoenzyme, isogenic, isoleucine, isomer, isopropanol, isotope
macro-	**makros** 'long, large': macrocephalic, macrolide, macromolecule, macrophage
mega-	**megas, megalo-** 'great': megabase, megabyte, megafauna, megalopolis
meso-	**mesos** 'middle, intermediate': mesoderm, mesophyll, mesothelium
meta-	**meta** 'with, across, after': metabolism, metadata, metamorphosis, metastasize
micro-	**mikros** 'small': microbe, microbiology, microgram, microscope, microtubule
nano-	**nanos** 'dwarf': unit of measurement: nanometre = 1 billionth of a metre, nanoparticle
neo-	**neos** 'new': neologism, neomycin, neon, neonate, neophyte, neoplasia, neoantigen
oligo-	**oligos** 'small', **oligoi** 'few': oligomer, oligonucleotide, oligopeptide, oligopoly
ortho-	**orthos** 'straight, right': orthodontics, orthodox, orthogonal
pan-	**pan** 'all': panacea, pancreas, pandemic, pantothenic acid
para-	**para** 'beside': para-aminobenzoic acid, paracrine, paradigm, paradox, paralogous
peri-	**peri** 'about, around': pericardium, perimeter, perinatal, period, peripheral, peristalsis

philo-	**philos** 'loving': philosophy. As a suffix: hydrophilic, electrophile, thermophile
poly-	**polus** 'much', **polloi** 'many': polyacrylamide, polymer, polynucleotide, polyploid
proto-	**protos** 'first': protein, proteome, protocol, proton, prototype, protozoa
pseudo-	**pseudos** 'falsehood': pseudogene, pseudomonas, pseudopodium, pseudouridine
pyr(o)-	**pur** 'fire': pyrogen, pyridine, pyridoxal, pyrimidine, pyruvic
syn-	**sun** 'with': synaesthesia, synapse, synchronous, syncytium, syndrome, synthesis
tachy-	**takhus** 'swift': tachycardia, tachypnea
tauto-	**tauto** 'the same': tautology, tautomers, tautonym
tele-	**tele** 'far off': teleconference, telemedicine, telephone, telescope
xeno-	**xenos** 'stranger': xenobiotic, xenograft, xenon, xenophobia, xenotransplant
xero-	**xeros** 'dry': xeroderma, xerography, xerophytes, Xerox

Table 7. Latin words and phrases that are used in English

Latin term	Literal translation	Meaning in English
AD (anno domini)	in year of the Lord	Placed after numerals to indicate years since Christ's birth, e.g. 1066 AD
ad hoc	to/for this	improvised or just for this
a fortiori	with strength	all the more so, with greater reason
ad infinitum	to infinity	endlessly, without limit
ad lib (ad libitum)	with freedom	freely, without restrictions
ad nauseam	to sea sickness	to the point of causing nausea
a priori	what is before	reasoning by deduction
a posteriori	what comes after	reasoning from experience
agenda	things to be done	things to be done
alma mater	bounteous mother	one's college or university
alter ego	intimate friend	intimate and trusted friend
alumnus	son, child, pupil	former pupil
appendix	supplement	external adjunct, addition
australis	southern	southern (as in aurora australis)
bona fide	with good faith	in good faith, with sincerity
campus	plain, grassland	university and its grounds
carpe diem	seize the day	make the most of the opportunity
caveat emptor	buyer beware	buyer beware
circa (ca. or c.)	around	about, approximately
compos mentis	of sound mind	of sound mind
confer (cf.)	compare	compare
con (contra)	against	against
consensus	agreement	agreement
corrigenda	items to be corrected	items requiring correction
curriculum vitae	course of one's life	CV, résumé
cursor	runner, courier	positional marker on screen
de facto	of fact	in reality, in practice
de novo	new	new, refreshed
emeritus/emerita)	honourable discharge	a mark of a distinguished career
et alii (et al.)	and others	and others (other authors)

etc. (et cetera)	and the rest	and so on
et seq. (et sequentes)	and the following	and the following
ex cathedra	from the chair	with authority
ex gratia	out of goodness	reward given without obligation
exempli gratia (e.g.)	for sake of example	for example, for instance
festina lente	hasten slowly	more haste less speed
honoris causa	for the sake of honour	for merit, not formal qualification
ibidem (ibid.)	in the same place	in the same place, article
id est (i.e.)	that is	that is, in other words
in absentia	in one's absence	an award made in one's absence
in extensor	in full	fully and entirely
in situ	in place	in its natural location
inter alia	among other things	among other things
in toto	in total	completely, wholly
in vitro	in glass	in a test tube, not in situ
in silico	a result obtained with a computer	
in vivo	in life	manipulation on a living organism
ipso facto	by that fact	as a consequence
memorandum est (memo)	to be remembered	to be remembered
modus operandi	way of working	method or process for task in hand
nil carborundum	Don't let the bastards wear you down (modern Latin)	
non sequitur	it does not follow	faulty reasoning
nota bene (NB)	note well	note well, take note
opus magnum	great work	an artist's or writer's major work
pass. (passim)	throughout	throughout
per annum	by the year	annually, (rate) for a year
per capita	by the head	for each person, individually
per centum	by the hundred	per cent, rate for a hundred
per diem	by the day	daily, day rate
per se	by itself	intrinsically, specifically
persona non grata	person not pleasing	unwelcome guest

post meridiem (p.m.)	after midday	afternoon, cf. ante meridiem
post mortem	after death	autopsy
post scriptum (PS)	after writing	footnote, note at bottom of letter
prima facie	at first sight	at first sight, initially
pro forma	for form	as a matter of formality
pro rata	by rate	proportionately
pro tempore (pro tem.)	temporarily	temporarily
quod erat demonstrandum (QED)	which was to be demonstrated	proved as required demonstrated
quod vide (q.v.)	which see	see for clarification
quorum	of whom	a specified minimum number
re	thing	concerning, regarding, relating to
scientia est potentia	knowledge is power	knowledge is power
sic	so, thus	calls attention to error in original
sine qua non	without which no(t)	indispensable, essential
status quo	situation in which	current situation, snafu (milit.)
sui generis	of its own kind	unique, original, distinctive
summa cum laude	with highest praise	with highest distinction
vide infra	see below	see below
vide supra	see above	see above
terra firma	firm land	solid earth
veto	I forbid	to disallow something
via	way	by way of
vice versa	turned position	complementary statement implied
viva voce	with living voice	oral exam
ultra	beyond	to an extreme degree

(Selected from www.businessballs.com/latin-terms-phrases.htm)

CHAPTER 3

Prefixes

This chapter enlarges on the lists of Latin and Greek prefixes in tables 5 and 6. Note that these prefixes can be attached to many different root words, as for **acro-** below. Compare with word elements such as **cyto-** or **osteo-**, also referred to as *combining forms*, which have restricted meanings, in these cases forming words to do with cells and bone respectively.

A: meaning 'not', or indicating the absence of something as in: *amorphous*, without a regular shape, or *alexia*, inability to read; *amnesia*, loss of memory, *apathy*, lack of interest. When the next letter is another vowel or an *h*, then a consonant, usually an *n*, is inserted, e.g. *anaerobic*, lack of air or oxygen, *anhydrous*, dry, i.e. without water. This convention derives from Greek. It also occurs in words derived from Latin where it modifies the *ad-* in words such as *ascend, aspire* as well as in OE words such as *afloat, aloud, awake, asleep*.

ACRO: **akros** (Gk.) 'at the end, tip, or top', hence *acropolis, acrobat*
An *acrocentric* chromosome is one where the *centromere* is located towards or near one end, while in a *telocentric* chromosome, the short arm is barely distinguishable. In a *metacentric* chromosome, the centromere is at or near the middle.
ACRAL, the adjective, in a medical context describes diseases such as gout and a rare form of melanoma that affects fingers and toes.
ACROMEGALY: **akros + megas, megal-** 'large, great'. A disease characterized by overdevelopment of the extremities, typically the result of overexpression of growth hormone by a pituitary gland tumour.
ACRONYM: **akros + onuma** 'name'. An abbreviation based on the first and sometimes additional letters of two or more words, as in ATP, DNA, PAGE (polyacrylamide gel electrophoresis). Among the literati, there is an opinion that a *true* acronym must be a pronounceable word such as *scuba*, *radar*, or *AIDS* and that everything else is simply an abbreviation. This line of thought can be safely consigned to the WPB. But while we are on this subject, it can be noted that there exists a type of reverse acronym where

all letters in an existing word are taken as the initial letters of other words. In the trenches during WWI, soldiers from Norwich decided that the name of their home town stood for 'Nightie off ready when I come home'.

ACROPHOBIA: **akros** + **phobos** 'fear'. Fear of heights.

ACROSOME: **akros** + **soma** 'body'. The anterior portion of the spermatozoon head, containing enzymes which function to penetrate the *zona pellucida* that surrounds the oocyte.

ACROSTIC: **akros** + **stikhos** 'row, line of verse'. Usually a word or words made up from the first or last letters of successive lines of a poem. These provide a temptation for pranksters and the disgruntled, so sub-editors keep a close eye out for acrostics. In 1961, an edition of the Australian weekly magazine *The Bulletin* was recalled when an acrostic was discovered, too late, consisting of the first letters of the lines of a submitted pair of sonnets, which read 'So Long, *Bulletin*, Fuck All Editors'. The author of this hoax was Gwen Harwood, a well-known poet, who was annoyed by having difficulty getting poems published under her own name but was having better luck when she used a male pseudonym.

Indenting alternate lines of a poem helps to camouflage an acrostic. With newspapers and magazines cutting back on sub-editors, a new golden age for the acrostically inclined could be dawning.

ALLO-: from **allos** (Gk.) meaning 'other' or 'different'

ALLOPATRIC: **allos** + **patra** 'fatherland'. Allopatric speciation occurs when members of a single ancestral species become geographically isolated; cf. *sympatric*: speciation which occurs when new species evolve while inhabiting the same geographical area.

ALLOPURINOL: 6-hydroxypurine. This was used as a treatment for gout, which results from crystals of sodium urate being laid down in joints. Urate is the end product of purine catabolism in humans and higher primates and is formed by the action of xanthine oxidase, which is inhibited by allopurinol.

See also entries for ALLOGRAFT and ALLOTROPY.

AMBI- (L.) 'on both sides' as in *ambidextrous*, meaning equally adept with either hand at tasks such as writing; from **dexter** (L.) 'right hand' or 'right side of the body'

AMBIENT: referring to the surroundings as in *ambient temperature*, i.e. the prevailing temperature; from **ambire** 'to go around' from **ambi-** + **ire** 'to go'.

AMBIGUOUS: open to more than one meaning or interpretation; from **ambigere** (L.) 'go around' from **ambi** + **agere** 'to do'.

AMBITION: desire for success, derived from the past participle of **ambire** (L.). AMBIVALENCE: having mixed feelings or contradictory attitudes or opinions, **ambi** + **valere** 'to be strong'. See under CONVALESCE for other words derived from **valere**.

AMPHI- (Gk.) meaning either 'on both sides', as in *amphibian*, **amphi-** + **bios** 'life', i.e. 'of both kinds', hence *amphipathic*: **amphi-** + **pathikos** from **pathos** 'experience' or *amphiphilic*, terms applied to molecules such as proteins and membrane lipids which contain both hydrophilic and hydrophobic regions.

AMPHIPODA: **amphi** + **pous, pod-** (Gk.) 'foot': an order of crustaceans, the name alluding to the possession of two types of legs, one specialized for swimming and the other for feeding.

AMPHIOXUS: **amphi** + **oxus** (Gk.) 'sharp', a *cephalocordate*, the name referring to its elongated anatomy, narrow or 'sharp' at both ends.

AMPHOTERIC: from **amphoteros**, 'both ways', used to describe a compound able to react both as an acid and a base, e.g. an amino acid

ANA- (Gk.) a prefix occurring in a very large number of Greek words where the sense implied by the prefix varies depending on context. Further variation in meaning has occurred since the adoption of these words into English. In a metabolic context, *anabolic* refers to *synthetic* reactions in contrast to *catabolic* or *degradative* reactions. In chemistry, *analogues* are compounds with related structure from **analogos**, meaning 'proportionate or equivalent to'; cf. *analogy*.

ANAPHYLAXIS refers to an extreme reaction, as in *anaphylactic shock*, resulting from exposure to an introduced antigen; from **ana-** 'again' + **phulaxis** 'guarding', with anaphylaxis indicating 'lack of protection'. Accordingly, *prophylaxis* indicates avoidance of disease by preventive treatment. See also under **ENZYMES** the circumstances surrounding the introduction of the terms *analysis* and *catalysis*.

ANTEPENULTIMATE: from Latin: **ante** 'before' + **paene** 'almost' + **ultimus** 'the end'. So antepenultimate means 'third from the end', e.g. the third amino acid from the end of a polypeptide chain. The *penultimate* amino acid in this case is the second last amino acid. The final amino acid could be called *ultimate* but normally is referred to as the C-terminal amino acid. The root **paene** is an adverb meaning 'almost', as in *peninsula* and *penumbra*. The final amino acid could be called *ultimate* but normally is referred to as the C-terminal amino acid.

ANTI- (Gk.) **anti-** 'against' or 'opposed to'.
ANTIDOTE: **antidoton**, literally 'given against'. A substance able to counteract the effects of a poison, an antidote may be specific, such as an antibody raised against the toxin, or non-specific, such as activated charcoal, which is able to adsorb many poisons and thereby prevent their effect.
ANTISEPTIC: **anti-** + **septikos** 'rotten'. Antiseptics are defined as antimicrobial substances that are applied to the skin or tissue, as distinct from *disinfectants*, which are applied to non-living surfaces. Antiseptics may be *bactericidal*, i.e. they kill bacteria, or *bacteriostatic*, in which case they inhibit their growth. Likewise, *asepsis* and *aseptic*, referring to conditions which exclude microorganisms. *Sepsis* thus describes poisoning due to bacterial infection and *septicaemia* the presence of microorganisms in the blood. See also **ANTIBIOTIC**.

APO- (Gk.) 'off, from, away'
APOPTOSIS: **apo-** + **ptosis** 'falling, a fall': the word introduced in 1972 to describe the phenomenon of *programmed cell death*, whereby some cells die as part of normal development of an organism.[10] Apoptosis was used by the Greeks to describe the falling off of petals from flowers or of leaves from trees in autumn. (The English word for this is *abscission*, from the Latin verb **abscindere** and from which the plant hormone *abscisic acid* takes its name.) Recent years have seen the introduction of the analogous words *necroptosis*, cell death due to injury or disease, i.e. *necrosis*, and *pyroptosis*, cell death occurring as a result of the inflammatory response caused by injury or infection and recently, *ferroptosis*, the death of an infected cell by iron starvation. APOCRINE: see under entry for **SECRETION MECHANISMS**.

AUTO- (Gk.) **autos** 'self, directed from within'

AUTOLYSIS: **autos-** + **lusis** from **luein** 'to loosen or split', refers to the self-digestion of cells by their own digestive enzymes released from *lysosomes*. This is usually confined to injured or dying cells.

AUTONOMIC: **autonomos** 'having its own laws' from **autos** + **nomos** 'law'. As in *autonomic nervous system*: that part of the nervous system that operates subconsciously to control functions such as heart rate and motility of the gastrointestinal tract.

AUTORADIOGRAPH: **autos** + **radius** 'ray' + **graphos** 'writing'. The pattern obtained when a mixture of radioactive components is subjected to a separation procedure such as gel electrophoresis and the gel then exposed to photographic film.

AUTOSOME: **autos** + **soma** 'body': any chromosome except a sex chromosome

See also AUTOCRINE, AUTOPHAGY.

BIO-: **bios** (Gk.) 'life' (Latin: **vita** as in vitamin)

BIOASSAY: **bios** + **assai** 'test' (OF): detection of a biological molecule by way of its biological activity.

BIOCHEMISTRY: the study of the chemistry of living organisms.

BIOCHEMISERY: repeating Biochemistry 101.

BIOENERGETICS: **bios** + **en-** 'within' + **ergon** 'work': the study of energy transformations in living systems

BIOETHICS: **bios** + **ethike** 'morals'. Ethics as it relates to medical and biological research.

BIOFILM: groups of cells or microorganisms that stick to each other and adhere to surfaces.

BIOINFORMATICS: **bios** + derivation of **informare** (L.) 'to shape, fashion, or describe'. The computer-based assembly and analysis of biological information, particularly relating to the huge amount of available DNA sequence information but also protein sequence and structure information.

BIOLOGY: **bios** + **-logia** 'subject or area of interest'. The study of all aspects of living organisms.

BIOPOLYMER: **bios** + **polus** 'many' + **meros** 'part'. Biological molecules such as DNA, RNA, proteins, and polysaccharides consisting of multiple monomer units.

BIOPSY: a small sample of tissue for analysis, from **bios** + **opsia**. In Greek, **opsia** was a combining form in nouns denoting a sight deficiency.

AUTOPSY, meaning **post-mortem** examination, illustrates how English derives from both classical languages.

BIOSPHERE: **bios** + **sphaira** 'ball'. The regions of the earth in which life is found.

BIOSYNTHESIS: **bios** + **sunthesis** (Gk.) from **suntithenai** 'place together'. Reactions that in living organisms form biological molecules.

BIOTECHNOLOGY: **bios** + **tekhne** 'art, craft' + -**logy** 'knowledge of'. Industrial-scale production of biologically active molecules such as hormones and antibiotics.

BRADY- (Gk.) **bradus** 'slow'.

BRADYCARDIA: **bradus** + **kardia** 'heart'. A slow heart rate, defined for adults as less than sixty beats per minute.

BRADYKINESIA: **bradus** + **kinein** 'to move': slow movement, which can be a symptom of Parkinson's disease.

BRADYPUS: **bradus** + **pus** 'foot' (as in octopus): genus of three-toed sloths, arboreal mammals of a sluggish nature. The sluggishness is partly explained by their rather low-energy diet; they are *folivores*, i.e. leaf eaters. Their unusually low body temperature, 33—34°C, probably also contributes to their slowness. See also XENARTHRA. Flanders and Swann had a song about these, the first verse of which goes:

> A Bradypus or Sloth am I, I live a life of ease,
> Contented not to do or die, but idle as I please.

CATA-: **kata-** (Gk.) 'down or downwards' (among a variety of other meanings) as in *catatonic, catabolism, catabolite, catalysis, cation*; often, but not always, the opposite of **ana-** 'up'.

CIRCUM- (L.) 'about, around', a prefix attached to a large number of Latin words, mainly verbs, and to a lesser extent in English: *circumcise, circumlocution, circumnavigate, circumspect, circumscribe, circumvent*. The Greek equivalent of **circum-** is **peri-**, and when combined with the related verbs, **ferre** in Latin, **pherein** in Greek, these give rise to the analogous words *circumference* and *periphery*.

DE- (L.) to remove, or indicating the absence of something

DEOXYRIBOSE and its numerous derivatives, i.e. the OH group at the 2 position of ribose is replaced by a hydrogen atom. See **NUCLEIC ACIDS** entry.

DEHYDRATE: **de-** + **(h)udros** (Gk.) 'water'. Removal of water.

DEHYDROGENASES: class of enzyme also referred to as oxido/reductases, in which hydrogen or reducing power is removed from a substrate and transferred to an acceptor coenzyme, which is usually NAD^+, $NADP^+$, FAD, or FMN. Enzymes that transfer hydrogen or reducing power directly to molecular oxygen are referred to as oxidases, e.g. CYTOCHROME OXIDASE—see under heading for **MITOCHONDRIA**. For the structures of the above coenzymes see chapter 5.

DIA- (Gk.) 'apart, through, in a line', as in **diametros** 'diameter'.

DIAGNOSIS: **diagnosis** from **dia** + **gnosis** 'knowledge'. A conclusion based on an understanding of appearances or symptoms of disease.

DIALOGUE: **dialogue** from **dia** + **legein** 'to speak'. Conversation, often exploring two different points of view.

DIALYSIS: **dialusis** from **dia** + **lusis** 'separating, loosening one from another'. Use of a semipermeable membrane to separate molecules based on their size difference, typically removal of salt from a protein solution by dialysis against water.

DIASTEREOISOMER: **dia** + **stereos** 'solid, hard, three-dimensional' + **iso-** 'the same as, like'. Diastereoisomers are stereoisomers, i.e. structural isomers that are not enantiomers. Enantiomers are pairs of stereoisomers that differ at all their stereocentres and are therefore mirror images of each other, e.g. D-glyceraldehyde and L-glyceraldehyde. Diastereoisomers that differ in configuration at only one stereocentre are EPIMERS, e.g. D-glucose and D-galactose. Also see entries for **CHIRALITY** and MONOSACCHARIDES under **CARBOHYDRATES**.

DIS- (L.): prefix indicating negation, absence, or removal, as in *disable, discard, discharge, disconnect, discontinuous, disinfect, disprove, dissociate*.

DYS- (Gk.): prefix indicating difficulty doing something or lack of function

DYSPHORIA: **dusphoros** 'hard to bear'. The opposite of euphoria.

DYSPLASIA: **dus** + **plasis** 'formation'. Disorder of the growth or development of cells or tissues.

DYSPEPSIA: impaired digestion, from Greek **duspeptos** 'difficult to digest'.

DYSPNOEA: **dus + pnoe** 'breathing'. Difficulty breathing, cf. sleep apnoea: temporarily stopping breathing while asleep.

DYSTOPIA: derived from *utopia* and meaning the opposite thereof, see **TOPO**- heading in Ch. 5.

DYSPROSIUM is element number 66 in the periodic table and is one of the fifteen chemically similar elements known as the lanthanides. Because of this similarity, it proved difficult to isolate and identify, so it was named Dysprosium by its discoverer after the Greek word **dysprositos** meaning 'hard to get'.

DYSREGULATE: disobeys the 'rule' which says that a Latin verb should go with a Latin prefix but, in the current literature, is usually favoured over *disregulate*. *Dysregulate* indicates that the system in question *is* regulated but abnormally so, and this is not the same as *deregulated* or *unregulated*. *Dysfunction* and *ferroptosis* also break this 'rule', and there are plenty of others. Medical dictionaries list lots of **dys**-type disorders.

ECTO-: **ek-** (Gk.) 'out', **ektos** 'outside'.

ECTODERM: **ektos + derma** 'skin'. The ectoderm is one of the three primary germ cell layers in the early embryo, along with *mesoderm* and *endoderm*. Differentiated tissues that derive from the ectoderm include *nervous* tissue and *epidermis*. See also entry for DERMA.

ECTOPARASITES: such as head lice and mosquitos and distinguished from ENDOPARASITES such as worms. See entry for **SYMBIOSIS**.

ECTOTHERMS: **ecto + thermos** 'heat'. These are organisms whose body temperature varies with that of the environment, i.e. cold-blooded animals such as amphibians, reptiles, and most fish, while *endotherms* are warm-blooded animals such as birds and mammals. Ectotherms used to be called *poikilotherms*, the *poikilo*- stem meaning 'variable', and endotherms were *homeotherms*, homeo- derived from **omoios** 'resembling' or 'of the same kind'.

EN- comes in various guises: from the Greek meaning 'within' or 'inside' but also as an OE prefix performing a variety of functions such as verb formation: *enhance, enable, encode*; as an OE suffix forming verbs from adjectives: *widen, dampen, tighten*; also forming past participles: *driven, forgotten, woken*.

ENDEMIC: **en-** + **demos** (Gk.) 'people'. Typically used in connection with a disease or condition such as malaria or malnutrition that is regularly prevalent within a particular area or population; see also *epidemic, pandemic* below.

ENDERGONIC and EXERGONIC: **end-** and **ex-** (Gk.) 'within' and 'out' + **ergon** 'work'. These terms refer to the free energy change (ΔG) for a reaction, i.e. the difference in free energy between reactants and products. An endergonic reaction is one in which ΔG is positive and will not occur spontaneously unless energy is supplied or by *coupling* it to an exergonic reaction. Conversely, an exergonic reaction *may* occur spontaneously when ΔG is negative. In general, synthetic reactions are endergonic and oxidation reactions are exergonic. See also the heading for **FREE ENERGY**.

ENDOGENOUS: **endo** + **genos** 'produced'. Literally 'produced from within', as opposed to EXOGENOUS. The cholesterol synthesized in the body is endogenous while dietary cholesterol is exogenous.

ENDONUCLEASE and EXONUCLEASE: **end-, ex-** + nuclease. Nucleases are enzymes that degrade nucleic acids. They may be *deoxyribonucleases* (DNases), which act on DNA, or *ribonucleases* (RNases), specific for RNA. Endonucleases cleave their substrates at internal positions; exonucleases act on terminal residues and may be specific for either 5' or 3' ends.

ENDOPEPTIDASE and EXOPEPTIDASE: enzymes that degrade polypeptides, i.e. proteins. Endopeptidases break internal peptide bonds, while exopeptidases liberate terminal amino acid residues and, depending on which terminus they recognize, are classified as *aminopeptidases* or *carboxypeptidases*.

ENDOTHERMIC and EXOTHERMIC: **endo-, exo-** + **thermos** 'heat'. These terms refer to reactions in which heat is absorbed or released. Dissolution of solid urea in water is an example of the former and the addition of concentrated sulphuric acid to water an example of the latter.

ENDOTOXIN: **endo-** + **toxikon** 'poison in which arrows were dipped', from **toxon** 'bow': a lipopolysaccharide found in the cell walls of Gram-negative bacteria which is recognized by the human innate immune system. See also ENDOCRINE, EXOCRINE under **SECRETION ECHANISMS** and ENDORPHINS under **RECEPTORS**.

EPI- (Gk.) prefix meaning 'upon, above, near, in addition'.

EPIDEMIC: **epi-** + **demos** 'people'. Defined as the increased occurrence of a disease compared to normal and is characteristic of *contagious* diseases such as flu. It is also used in connection with lifestyle diseases such as type 2 diabetes. See also ENDEMIC above and PANDEMIC below.

EPIDEMIOLOGY: **epidemia** 'prevalence of disease'. The branch of medicine that studies disease prevalence, their distribution, and environmental factors that may contribute to particular diseases.

EPIDERMIS: **epi** + **derma** 'skin'. The outermost layer of cells in both plants and animals. Related words: *dermatitis, dermatology, intradermal, hypodermic* (syringe), *epidermal growth factor*. See also entry for **DERMA**.

EPIMER: a stereoisomer that differs in configuration at only one stereocentre, e.g. D-glucose and D-galactose or D-glucose and D-mannose.

EPINEPHRINE: **epi** + **nephros** 'kidney'. Hormone secreted by the medulla of the adrenal glands, which are situated, as the names of the hormone and the glands suggest, above the kidneys. It is also known as adrenalin.

EPIPHYTE: **epi-** 'in addition' + **phuton** 'plant'. A plant that grows on another plant but is not parasitic, such as orchids that grow on the bark of rainforest trees.

EPITHELIUM: **epi** 'on or upon' + **thele** 'breast'. *Epithelial tissues* are one of the four basic types of animal tissue, the others being *connective tissue, muscle tissue,* and *nervous tissue*. Epithelial tissue is specialized to form the covering or lining of all internal and external body surfaces. Epithelial cells also form many glands, both *exocrine* and *endocrine*. In English, the **thele** root has been applied to tissue in general, thus *endothelium, mesothelium*. See also entry for **TISSUE**.

EPITOPE: **epi** + **topos** 'place': a region in a molecule such as a protein where a particular antibody will bind, i.e. an *antigenic determinant*. See also entry for EPIGENETICS.

EXTRA- (L.) 'outside or beyond' as in *extracellular, extracurricular, extrapolate*; cf. INTRA-.

GENE, GEN-, and -GEN: **genos** (Gk.) 'race, kind or offspring'. The word *gene*, as a noun, occurs in multiple contexts and therefore is impossible to define precisely. The same stem gives rise to the prefix *gen-* as in *genetics, genotype, gender,* and *genealogy*, while the suffix *-gen* indicates the production of something, as in *hydrogen, nitrogen, pathogen, immunogen*, etc.

GIGA-: from **gigas** (Gk.) 'giant'. A combining form indicating 10^9, i.e. one billion or a thousand million, as when the sizes of genomes are expressed in *gigabase pairs*, also *gigabytes* of download etc. See also table 10.

HETERO- (Gk.) 'the other (of two), other, different'.
HETERODUPLEX: **hetero-** + **duplex** (L.) 'having two parts'. Used in reference to a double-stranded nucleic acid where the strands are different, as in a DNA-RNA hybrid.
HETEROGENEOUS: **hetero-** + **genos** 'sort, kind'. It usually refers to something that is a mixture. The word is perhaps best defined as the opposite of homogeneous. One also occasionally comes across *heterogenous*, which the *OED2* describes as 'a less correct form of heterogeneous'. According to the *Cambridge Guide to English Usage*, *heterogenous* is used by biologists in the specialized sense of 'from outside the body, of foreign origin', which is unfortunate since it is bound to be confused with *heterogeneous*. Forget *heterogenous*! (I'm sorry I mentioned it.) See also *homodimer* and *heterodimer* below.

HOLO-: **holos** (Gk.) 'whole, entire'
HOLOENZYME: a term typically used in connection with enzymes that carry out complicated reactions, or reactions subject to stringent regulation, such as DNA and RNA polymerases, and composed of multiple subunits required for their *in vivo* functions. Their isolated catalytic subunits often display catalytic activity *in vitro*.
HOLOCAUST: **holo-** + **kaustos** 'burned'. Slaughter on a mass scale.
HOLOCENE: **holo-** + **kainos** 'new': the current geological epoch beginning about 10,000 years ago, corresponding to the end of the last ice age. See also under ANTHROPOCENE.

HOMEO-: **omoios** (Gk.) 'resembling' or 'of the same kind'
HOMEOSTASIS: **omoios** + **stasis** 'standing'. The French physiologist Claude Bernard in the mid-nineteenth century introduced the idea that good health involved regulation of all the bodily systems in order to maintain the *milieu interieur* what is now referred to as homeostasis. Everything that has been discovered since Bernard's day about how living systems work has confirmed this idea, and much biological research has been directed towards understanding how this regulation is achieved.

Familiar terms are *immune homeostasis, glucose homeostasis, caloric homeostasis, protein homeostasis.* See also HOMEOPATHY under **PATHO.**

HOMO-: **homos** (Gk.) 'the same'
HOMODIMER: describes a protein which functions as a dimer composed of two identical subunits, as distinct from a *heterodimer.*
HOMOGENATE: the result of the mechanical disruption or homogenizing of tissue, the starting point for most biochemical preparations
HOMOGENEOUS: **homos + genos** 'race or kind'. A true solution is homogeneous, as distinct from a suspension. The term also applies, for example, to a purified protein preparation which contains a single component.
HOMOZYGOUS: **homo- + zygotos** 'joined'. An organism having identical alleles, one from each parent, at a particular genetic locus, cf. *heterozygous.*

HYPER- (Gk.): 'above, exceeding, beyond'. Corresponds to Latin **super-**
HYPERGLYCAEMIA: **huper- + glyco-** 'to do with sugar' + **-aemia** 'blood': excess blood glucose, usually a sign of diabetes.
HYPERPLASIA: **huper- + plasis** 'formation'. Enlargement of an organ or tissue due to increase in the *number of cells.*
HYPERTROPHY: **huper- + trophe** 'nourishment'. Enlargement of an organ or tissue due to increase in the *size of its cells.* Common usage is degrading the distinction between *hyperplasia* and *hypertrophy*, as in 'the hypertrophy of our frontal lobes'. See also *hypertonic* below under ISOTONIC heading.

HYPO- (Gk.): 'under, below, beneath'.
HYPOTHESIS: **hupothesis**: from **huper + thesis** 'placing', which meant 'foundation' in Greek but a proposed explanation for an observation or phenomenon in English
HYPOXIA: Lack of oxygen. See also *hypotonic* below under **ISO-.**

IN-: a Latin prefix. English words that contain the **in-** prefix usually derive from Latin. **In-** as a prefix can have two main meanings: either 'not' or 'in or into'. Examples meaning not, or at least having a negative connotation: *inoperable,* meaning in a medical context, not able to be operated upon;

Inorganic: in chemistry, means 'does not contain carbon', i.e. is 'not organic'; when it comes to organic food, the opposite would probably be 'non-organic'.

Inhibit/inhibition—as used in connection with enzyme inhibition; the Latin verb is **inhibere**, *ppp.* **inhibitum**, meaning 'to check or hinder'. **Inhibere** comes from **in-** + **habere**, 'to hold', so there is no English word 'to hibit'. Likewise, words such as *innocent, injury, innate, initiate*, etc. For most of the hundreds of English words where **in-** means 'not', the **in-** could be replaced by *not* as in *infinite, insoluble, invisible*. An example where **in-** means the opposite of 'not' is *inflame* and *inflammable*, the latter meaning 'flammable' or 'able to burn', is a problem since it is routinely used in safety and warning notices where it could be misunderstood to mean its exact opposite, namely 'non-flammable'. The only other similar example is *incandescent*, meaning 'glowing white hot' and which means the same as the obsolescent *candescent*. Words where **in-** means 'in' or 'into' include *include, inscribe*, and *inoculate*. Finally, there are a large number of words where, depending on the first letter of the root word, **in-** becomes **im-** as in *imbibe, implant*, or **il-** as in *illegible, illuminate*, and *illustrate*. See note on 'assimilation' in chapter 2.

INFRA- (L.) 'below' as in *infrastructure* and *infrared*, that part of the electromagnetic spectrum in the wavelength range 700 nm–1 mm. It occasionally shows up in official documents as **vide infra**, 'see below', while *infra dig* may describe behaviour lacking dignity. Cf. *supra*.

INTER-: from the Latin prefix **inter-** meaning 'between' or 'among' and, usually, the same in English, as in *interfere, intermolecular, international, intercellular, interdisciplinary, interface, interpolate*. As with the prefix **in-**, **inter-** can be subject to assimilation: *intelligence* and *intellect* derive from **intellegere, intellexi, intellectum**: 'to perceive, understand'.

INTERCALATE: from **intercalarius** (L.) 'to be inserted'. A term used to describe the ability of hydrophobic planar molecules such as *acridine* and *ethidium bromide* to insert themselves between DNA base pairs. The word originates from the time when it was necessary to insert extra days, *intercalary days*, in early versions of the calendar in order to keep it aligned with the seasons. The 29th of February, which occurs every four years, is intercalated in the current calendar.

INTERFERONS are a group of proteins that are produced by cells in response to infection by bacteria and viruses and named on account of their ability to interfere with viral replication. They activate the immune system and initiate signalling pathways that enable neighbouring uninfected cells to resist infection. INTERLEUKINS: a large group of *cytokines* or signalling proteins, also referred to as lymphokines, first detected as secretion products of *leukocytes*, i.e. white blood cells, but now known to be synthesized by a variety of cells. Their functions include regulation of the immune system.

INTRA- (L.): prefix meaning 'within' or 'on the inside', thus *intramolecular, intracellular, intrauterine*, etc. Often in contrast to *extra-*.

ISO- **isos** (Gk.) 'equal to', 'the same as', 'like'

ISOMERS: **iso** + **meros** 'part'. These include structural isomers where two molecules have the same formula but a different arrangement of chemical groups. Examples are the amino acids leucine and isoleucine, also the glycolytic intermediates dihydroxyacetone phosphate and glyceraldehyde 3-phosphate. Enzymes able to interconvert isomers are *isomerases*. See also under the headings **CHIRALITY** and MONOSACCHARIDES under **CARBOHYDRATES**.

ISOELECTRIC POINT: the pH at which a protein contains equal numbers of positive and negative charges and therefore does not migrate in an electric field.

ISOPEPTIDE BOND: This results from bond formation between the C-terminal carboxyl group of a protein and the ε-amino group of a lysine residue in another protein. Proteins marked for degradation are linked via isopeptide bonds to the carboxyl group of *ubiquitin*. See also entries for PROTEIN DEGRADATION under **PROTEINS** and BLOOD CLOTTING under **BLOOD**

ISOTONIC: **tonos** 'tone, strength, intensity'. An isotonic solution, e.g. 0.9 per cent (w/v) NaCl, is one that will not cause osmotic volume changes in cells suspended in it. Cells suspended in a *hypertonic* salt solution, i.e. more concentrated than isotonic, will shrink because of the loss of intracellular water. Cells suspended in a *hypotonic* solution will *imbibe* water and eventually *lyse*.

JUXTA-: **iuxta** (L.) 'nearby, close' as in *juxtaposition*: in close proximity, next to or side by side.

MACRO-: **makros** (Gk.) 'long', as in *macromolecule, macrostructure, macroscopic, macrocephalic,* etc. See also MACROPHAGES under **PHAGE**.

MAL-: **malus** (L.) 'bad, harmful', as in *malabsorption, malady, malaise, malfunction, malnutrition, malodorous. Malaria* comes from the Italian **mala aria** 'bad air', since originally its cause was attributed to the 'bad air' of putrescent organic matter emanating from swamps and marshes. Such a noxious vapour was known as a *miasma,* from the Greek meaning 'stain', and in earlier days was the alleged cause of various diseases.

MEGA- (Gk.): 'large'. SI prefix indicating a million (10^6), symbol M, as in *megabase pairs, megadaltons, megabytes,* etc. See also table 10.
MEGA-, MEGALO-: as in *megalopolis*: very large city; *megalomania*: exhibited by people with an obsession to dominate others. As a suffix: *hepatomegaly,* enlarged liver; *splenomegaly,* enlarged spleen etc.

META-: **meta**. In Greek an all-purpose preposition and adverb with more than twenty meanings depending on context. In English an equally all-purpose prefix. The Macquarie Dictionary gives the following meanings: 'among, together, with, after, behind, along with'. Among scientific or technical words, the most common senses of *meta-* imply change or multiplicity.
META-ANALYSIS: the deployment of statistical methods on multiple research results to consolidate conclusions from individual studies.
METASTABLE: **meta- + stabilis** (L.) from **stare** 'to stand' or 'steady, stable'. A description applied to systems or chemical or physical states that are intrinsically unstable but nevertheless may remain in the metastable state indefinitely. In many cases, transition to the more stable or ground state is the result of a minor perturbation. An example is *supercooling* of highly purified water below its normal 0°C freezing point. Introduction of a minute impurity can then *nucleate* ice formation. Similarly, on account of their small size, water droplets in clouds have to be colder than -15°C before they freeze. Even then, a dust particle is usually needed to seed ice formation. Another example of a metastable system is, as seen on TV from time to time, an assembly of several thousand dominos standing on edge. When the first domino falls, so do the rest in sequence. See Ball (2001) on the various metastable forms of ice. See also **METABOLISM** and metastasis under **CANCER**.

MICRO-: **mikros** (Gk.) 'small'. An SI prefix indicating 10^{-6} or one millionth, symbol μ. Thus, 1 μg = 1 millionth of a gram. See also table 10. Micro also serves as general prefix meaning small as in *microbe*, an organism only visible under a *microscope*: **mikros** + **skopein** 'to look at', hence *microscopic*, cf. *macroscopic*: visible to the naked eye. See also entries for CRYO-ELECTRON MICROSCOPY and -**SCOPE**.

MULTI-: **multus** (L.) 'much, many' as in *multiple, multiply, multicellular, multifunctional, multifactorial*. See the entry for **STEM CELLS** for the meanings of *multipotent* and *pluripotent*.

NANO-: **nanos** (Gk.) 'dwarf'. SI prefix, indicating a factor of 10^{-9} as in nanosecond, nanometre, etc., symbol n. Nano- is also used to indicate very small particles and processes as in *nanoparticle, nanostructure, nanotechnology*. See also table 10.

NEO-: **neos** (Gk.) 'young or new'. So we have the inert gas, *neon*, which was new in 1898 when it was discovered; *neonatal*, meaning newborn; *neologism*, a newly coined word or expression.
NEOPLASM: **neo** + **plasma** 'formation'. Abnormal new growth.
NEOTENY: **neos** + **teinein** 'to extend': retention of juvenile characteristics in the adult, also known as *paedomorphism*: **paedo-** 'child' + **morph-** 'form'.

OLIGO-: **oligos** (Gk.) 'few or small', thus *oligomer, oligonucleotide, oligopeptide, oligosaccharide*. The point at which an oligonucleotide or oligopeptide becomes a polynucleotide or polypeptide is ill-defined.

OMNI-: **omnis** (L.) 'all' as in *omniscient*, **scire** 'to know', hence to know everything; *omnivorous* from **vorare**: 'to swallow', which describes people and other animals who are not fussy about their food. *Devour* comes via OF from the same root.

PAN-: **pan** (Gk.) 'all' as in *pandemic*, **pan-** + **demos** 'people': an epidemic disease occurring over a wide area, e.g. the worldwide flu epidemic in 1918 and COVID in the 2020s. See *epidemic* and *endemic* above.
PANACEA: **panakeia**: universal remedy.
PANCREAS: **pankreas** from **pan** 'all' + **kreas** 'flesh'. Here **pan** serves to emphasize the fleshiness of the gland, also known as sweetbread in a

culinary context, which secretes insulin from cells known as the islets of Langerhans, from which insulin derives its name (**insula** L. 'island, isle'). The pancreas also secretes a range of digestive enzymes into the stomach.

PANGAEA: **pan** + **gaia** 'earth', postulated early supercontinent that later fragmented

PANTOTHENIC ACID: **pantothen** 'from all quarters, from every side'—alluding to its widespread occurrence; it is a component and precursor of coenzyme A. Sometimes referred to as vitamin B5.

PARA- (Gk.) 'beside or near'.

PARALLEL: **parallelos** from **para-** 'alongside' + **allelos** 'one another'. The strands of double-stranded DNA are said to be *antiparallel* because they have polarity and run in opposite directions.

PARATHYROID GLAND: a gland that regulates calcium levels and bone metabolism, located near the thyroid gland.

PAROTID GLAND: **para-** + **ous, otos** 'ear', one of the salivary glands, situated just in front of the ear with a duct that runs forward across the cheek and empties into the back of the mouth.

PARA- has acquired an additional niche meaning: *to protect*. This arose originally in Italy, where *parasol* came to mean protection from the sun; later in France *parapluie* (**pluie** F. 'rain') meaning umbrella, i.e. protection from rain, and then *parachute* (**chute** F. 'fall'), invented during WWI with the beginning of aerial combat.

PER (L.): 'through, all over, complete'. In Latin, it was used both as a preposition and as a prefix. Several examples of the former are used routinely without translation: *per annum, per capita, per diem,* and *per se,* while *per cent,* being an abbreviation of *per centum,* is regarded as sufficiently anglicized to be written without a space after the *per.* In chemistry, *per-* may indicate compounds containing the highest proportion of oxygen: thus, we have *hydrogen peroxide,* H_2O_2, with an extra oxygen compared to water, also the oxidizing agents *perchlorate,* ClO_4^-, *permanganate* MnO_4^-, *periodate* IO_4^-. *Peroxidases* are enzymes that catalyse reactions in which an alkyl peroxide is reduced to water and an alcohol in the presence of a reductant such as ascorbic acid:

$$RCOOH + AH_2 \rightarrow RCOH + H_2O + A.$$

(The COOH here is not a carboxyl group, rather C-O-O-H, a peroxide.) An unusual peroxidase, part of the defence against bacteria, is *myeloperoxidase*, so named because it was first isolated from the blood of patients suffering from myeloid leukaemia (**muelos** Gk. 'marrow'). It is present in the white blood cells, neutrophils, and monocytes, where it catalyses the reaction $H_2O_2 + 2Cl^- \rightarrow 2HOCl$. The product *hypochlorous acid* is strongly bactericidal. See also entry for LEUKOCYTES under **BLOOD**.

Other words with the per- prefix are *perpendicular, permeable, perfusion*, the latter pertaining to the supply of liquid to tissues, organs or bodies, situations where either *perfusion* or *infusion* could probably be used without causing *confusion. Pertussis* is another name for whooping cough (**per-** 'extremely' + **tussis** 'cough'). It seems a certain degree of etiquette is in order when it comes to the word *perspire*. I remember being told as a child that people don't sweat—horses sweat, but people *perspire*. I am now reliably informed that ladies no longer perspire—they *glow*, as in the line 'Women glow and men thunder.' The word *perfect* (from **perfectus**, the past participle of **perficere** 'to complete') can be used in a phrase such as *a perfect shambles*, i.e. a complete shambles, while in grammar, the *perfect tense* indicates an action that is past or completed.

PERI-: (Gk.) prefix meaning 'around or about'. Frequently occurs in words describing membranes, fluids, or structures that surround other structures, e.g. *periosteum*, the vascular connective tissue that envelops bones; *pericardium* (**kardia** 'heart'), the membrane enclosing the heart; *periodontics*, concerning the gums, membranes, etc. surrounding teeth (**odous, odont-**). *Peritoneum* from **peritonos** 'stretched around' is the membrane lining the abdominal cavity. *Perinatal* from **nascor** (L.) 'to be born' means the time before and after birth. *Peristalsis*, from **stalsis** 'contraction', describes the rhythmic contractions that result in passage of food through the alimentary canal, also *peristaltic pump*, which works the same way. In Gram-negative bacteria, the space between their inner and outer membranes is known as the *periplasm*, **plasma** 'substance', cf. *cytoplasm*. The *peripheral* (**pherein** Gk. 'to bear') nervous system refers to the nervous system outside the brain and spinal cord, while *peripheral* blood refers to blood that is beyond the bone marrow, its site of synthesis. Not quite so obviously, *period* comes via OF from **peri-** + **hodos** the same root in *method, anode*, etc., hence the *periodic table*. **Hodos**, or **odos** in Greek without the *h*, translates variously as 'way' or 'a way of doing things' or perhaps 'system'. See entry for ELECTRODE under **ELECTRO-** for further words based on the **hodos** root.

PHILO-: **philos** (Gk.) 'loving', can serve as both prefix as in *philosophy*, **sophos**: 'wise', but more commonly as a suffix: *hydrophilic*, meaning polar and therefore able to interact with water; some bacteria are described as *thermophilic*, literally 'heat loving', i.e. heat tolerant. Some *Archaea* are *very* heat tolerant and are referred to as *extremophiles*.

PICO-: Spanish origin, meaning small: unit of measurement for 10^{-12} or 1 million millionth, symbol p. See also table 10. There is also a group of small RNA viruses termed picornaviruses, which includes polio, hepatitis A, and the animal virus foot-and-mouth.

POLY-: **polus** (Gk.) 'many' or 'much'. Cf. **polloi** 'many' as in 'hoi polloi'. A key word here is POLYMER, so depending on what sort of polymer you wish to specify, it can be *polynucleotide*, *polypeptide*, *polysaccharide*, etc. Enzymes that synthesize polymers are *polymerases*, although usage confines this to the various RNA and DNA polymerases.
POLYPLOID refers to the presence of more than the standard two sets of chromosomes, e.g. in wheat where different varieties are either *tetraploid* or *hexaploid*. This can also be a feature of particular organs, e.g. the liver, where some cells are *polyploid*. See also under **PLOIDY**.
POLYTENE (**tainia**: 'band') is the term applied to the giant chromosomes found in *Drosophila* salivary gland cells which consist of many *chromatids* in parallel alignment.

PRE-: **prae-** (L.) 'before' in time, place, or importance.
PREBIOTIC: refers to conditions on the earth before the appearance of life.
PRECIPITATE: **prae-** + **caput** 'head', in its familiar chemistry context as a verb, means to treat a solution so as to cause the *solute*, or portion thereof to become insoluble and settle out. To precipitate originally had the meaning 'to throw down headlong'.
PRECIPITIN: the insoluble precipitate formed by reaction of an *antibody* with an *antigen*, often visualized as a line on an agar gel.

PRECURSOR: **prae** + **currere** 'to run', a compound or metabolite from which another is formed.

PREDISPOSITION: **prae** + **disponere**: 'to arrange': a tendency, which can be genetic or psychological, to suffer from or be more likely to develop a particular condition, whether benign or harmful.

PRO- (Gk.): prefix meaning 'before' in various contexts as for PRE- above.

PROBOSCIS: **pro** + **boskein** 'to feed': this was the word the Greeks and Romans used to refer to the elephant's trunk. More recently, apart from serving as a humorous reference to a person's nose, it describes the feeding mouth parts of insects such as mosquitoes.

PROGERIA: **progeros** 'prematurely old', from **pro** + **geras** 'old age'.

PROKARYOTE: **pro** + **karuon** 'nut or kernel'. These are the microorganisms that do not contain a membrane-bound nucleus and consist of the two life domains *Eubacteria* and *Archaea*. See also **ARCHAEA** entry.

PROPHAGE: **pro** + **phagein** 'to eat': a bacterial virus inserted into its host chromosome. See also entry for **PHAGE**.

PROTO-, from **protos** (Gk.): 'first or foremost'. As well as being the root for *proton* and *protein*, where the sense was 'first' and 'of first importance', *proto-* is sometimes used to name a postulated precursor, hence *proto-ribosome, protocell*, etc.

PROTOCOL: a detailed procedure for performing an assay or experiment

PROTOTYPE: the first model, e.g. of a driverless car, that will be subject to further development

PROTOZOA: the group of phyla containing single-cell eukaryotes belonging to the kingdom PROTISTA

RE- (L.): prefix meaning 'once more or again', which can be attached to most verbs in the English language to indicate repetition. *Re*, without the hyphen, also is used meaning 'concerning' as in 're the meeting'. In this case, it is derived from the ablative of **res** (L.): *thing, matter, event*.

SEMI- (L.) half, as in *semiconservative replication of DNA*: referring to the fact that when a double-stranded molecule of DNA is replicated to produce two daughter molecules, each daughter contains one strand from the original molecule and one strand that is newly synthesized. The original molecule is therefore *semi-conserved*. Other examples of semi-: *semi-conscious* and *semi-synthetic* that describes naturally occurring antibiotics that have been further modified. The equivalent in Greek is **hemi-** as in *hemiacetal*, the product of the reaction between an aldehyde and an alcohol; a familiar intramolecular example is the cyclization of the open-chain form of glucose and other *aldohexoses* to form the intramolecular *pyranose hemiacetals*. A *hemiketal* results from the reaction for fructose of its keto group and an -OH group to form *fructofuranose*. See also under **CARBOHYDRATES**.

SUB- (L.): 'under or close to', as in *subcellular, subspecies, sub-zero*

SUPER- (L.): 'above' or 'beyond', as in *supernatant* from **super + natare** 'to swim' and *superbugs*, journalists' name for bacteria resistant to most antibiotics, also *superficial, superfluous, superior, superlative, supervise*
SUPERCOILED DNA: Many DNA molecules from viruses and bacteria are circular and supercoiled. Simply joining the two ends of a linear DNA molecule will generate what is termed a *relaxed circle*. If instead the DNA is caused to unwind corresponding to a few turns of the helix and *then* the ends are joined, the unwound DNA rewinds, and because the ends are no longer able to rotate with respect to each other, the result is a supercoiled molecule. Supercoiling of DNA causes it to become more compact. DNAs with varying numbers of supercoils are termed *topoisomers*. See also entry for **TOPO-**.

SUPRA- (L.): 'above, beyond', as in *suprarenal*: located above the kidneys. SUPRAMOLECULAR: of a structure with many molecules, e.g. ribosomes, virus particles.

SYN-: prefix meaning 'with' or 'together' in Greek and with the meaning in English of 'acting together' or 'bringing together'.
SYMPTOM: **ptoma**: fall, so **symptoma**: a falling together.
SYNAPSE: **syn- + hapsis** 'joining': the junction of two nerve cells, separated by a narrow gap or synaptic cleft, across which *neurotransmitters* pass when the *presynaptic cell* signals to the *postsynaptic* cell.
SYNDROME: **syn + dromos**: 'running': the concurrence or 'running together' of symptoms. In medicine, it refers to the association of a number of clinically recognizable features that, together, are characteristic of a particular condition.

SYNERGY or SYNERGISM: **sunergos** 'working together' from **sun** + **ergon** 'work'. This is where the combination of organizations, systems, or substances such as drugs produces a combined effect greater than a simple additive effect.

SYNOPSIS: **sun** + **opsis** 'aspect or appearance', summary or outline

SYNTENY: **sun** + **tainia** 'band' or 'ribbon', same root as in *polytene*—see above. Indicates similar gene arrangements on chromosomes or parts thereof from different species.

SYNTHESIS: **sun** + **tithenai** 'to place', hence 'place together', hence *synthetic, synthesize, photosynthesis*, etc. Enzymes catalysing synthetic reactions are *synthases* or *synthetases*. See also entries for SYNDROME and SYMBIOSIS.

TAUTO-: *tautomer*: **tauto** (Gk.) 'the same' + **-meros** 'part'. A tautomer is one of two or more *isomers* that are in equilibrium. This usually involves the migration of hydrogens, as in *keto-enol, quinine-quinol*, and *amino-imino* tautomerism. In 1953, Watson and Crick were closing in on a structure for DNA. As Watson (1970) relates in *The Double Helix*, he was trying, using cardboard models, to see if bases were able to pair via hydrogen bonds. This led nowhere until his colleague Jerry Donohue pointed out that the models for thymine and guanine he was using were the *enol* forms, as illustrated in the textbooks of the time, and that the dominant tautomer in both cases would almost certainly be the *keto* form. Accepting this advice, Watson went back to his cardboard models, and the rest is history. When DNA replicates, it does so amazingly accurately. Nevertheless, mutations do occur, and it has been suggested that the incorporation of the minor enol forms leading to mispairing of bases may contribute to this.

TAUTOLOGY: **tauto-** + **-logos** 'speech': saying the same thing twice in different words, as in 'a capacity crowd completely filled the stadium' or 'round circle' or 'Everyone agreed, and the vote was unanimous.'

TAUTONYM: **tauto-** + **onuma** 'name': a scientific name in which the same word is used for both genus and species, e.g. *Rattus rattus*, the black rat.

TELE- **telos** (Gk.) 'end' or 'at a distance'.

TELESCOPE: **tele-** + **skopein** 'to look at'. The idea here of **tele-** denoting something at a distance is also present in *telephone* and *television*. The **telos** root also occurs in words used to classify how different organisms deal with excretion of excess nitrogen, i.e. the end-point of nitrogen catabolism. Most terrestrial vertebrates excrete nitrogen in the form of urea and are UREOTELIC, birds and reptiles as uric acid and are URICOTELIC, while fish excrete it directly as ammonia and are said to be AMMONOTELIC.

TELOCENTRIC: **telo- + centrum** (L.), chromosomes in which the centromere is located very close to one end.

TELOMERES: **telo- + meros** 'part'. Special DNA sequences at the ends of eukaryotic chromosomes. TELOMERIC: The orientation of genes on chromosomes can be specified as either *centromeric* or *telomeric*.

TELOMERASE: enzyme containing both protein and RNA, which is essential for maintenance and replication of telomeres. See also entry for **GENOMES**.

TELOPHASE: the period, at the end of the cell division cycle, when new nuclear membranes are formed.

TRANS-: Latin prefix meaning 'across'. In chemistry, a distinction is made between the relative orientation of groups across a double bond, either *cis* or *trans* to yield stereoisomers. See **FATTY ACIDS**.

TRANSFERASES constitute a very large group of enzymes such as the aminotransferases that carry out reactions between amino acids and keto acids, acyltransferases as in fatty acid synthesis and phosphotransferases that phosphorylate substrates.

TRANSITIONS and TRANSVERSIONS: mutations in DNA are classified as transitions when one pyrimidine is replaced by the other pyrimidine or a purine is replaced by the other purine. Transversions are when a pyrimidine is replaced by a purine or a purine by a pyrimidine. See **NUCLEIC ACIDS and GENE EXPRESSION** for TRANSPOSONS, TRANSCRIPTION, and TRANSLATION.

ULTRA- (L.): all-purpose prefix with the sense 'extreme' as in *ultramarathon*.

ULTRACENTRIFUGE: high-speed centrifuge, both *preparative* and *analytical*.

ULTRAFILTRATION: filtration that removes particles in the size range of bacteria or smaller.

ULTRASTRUCTURE: biological structures at the electron-microscope level

ULTRAVIOLET: the short-wavelength region beyond the visible spectrum.

UN-: Old English prefix and often occurs with OE root words such as *unearth*, *unfair*, *unfeeling*, and *unhealthy*. In some words, the sense is of reversal, such as *undo*, *uncover*, *unfold*. **Un-** also occurs instead of **in-** in words such as *uninterested* or *unintelligible*, where the meaning is negative but *ininterested* etc. would not sound right. Notably, *uninterested* and *disinterested* are not synonyms.

CHAPTER 4

Short Etymologies

ABSTRACT: from **trahere, tractum** (L.), 'to draw together': *abstract, contract, detract, distract, extract, traction, intractable, retract, subtract*

ACIDIFY: to make or cause to become acidic. This is an example of a large number of words ending in *-fy* that derive from French words ending in **-fier** or have been invented by analogy. All these words that include the idea to make or to cause derive ultimately from the Latin ending **-ficare** and the verb **facere** 'to make'. Other examples that come to mind in the current context include *amplify, clarify, classify, emulsify, exemplify, falsify, humidify, identify, justify, liquefy, magnify, modify, purify, quantify, saponify, satisfy, simplify, solidify, specify, verify*.

ACTION: from **agere, actum** (L.), 'to lead, guide': *agency, act, exact, activate, interact, inactive, proactive, react, reagent, retroactive, radioactive, transaction*. Also **agonia** (L.): *contest, agony*; **agein** (Gk.); 'to lead, guide', *promote, stimulate*, hence *agonist, antagonist*; also **-agogue** (Gk.) from **agogos**: 'leading, guiding, attracting' as in *synagogue, pedagogue, demagogue, galactagogue, sialogogue*, these last two referring to stimulation of the secretion of milk and saliva respectively.

ACUTE: from **acutus** 'sharp, pointed', from **acuere** 'to sharpen' and **acus** 'needle': in the context of disease, meaning 'to come to a sharp crisis', as distinct from chronic; *acuity*, as in 'acuity of vision'; also, *acumen, acupuncture*.

ADHERE: from **haerere** (L.) 'to stick to', also *cohere, cohesive, inherent*. Also meaning to adhere, glue, or fasten is **agglutinare** (L.) derived from **gluten** (L.) 'glue'. An *agglutinin* is any substance that causes clumping of cells. *Gluten* in English refers to a protein fraction in cereal grains that is responsible for *coeliac disease* (**koilia** Gk. 'belly'), an autoimmune condition also referred to as *gluten enteropathy*. *Glutamine* was originally isolated from gluten.

AEROBIC, ANAEROBIC: **aer-** (Gk.) 'air' + **bios** 'life'. These terms were introduced by Pasteur in 1863, **aerobie** in French. They signify whether an organism such as yeast requires oxygen to survive. A *facultative* anaerobe, as distinct from an *obligate* anaerobe, generates ATP by aerobic *respiration* when oxygen is present, but switches to *fermentation* under anaerobic conditions.

AETIOLOGY: **aitios** (Gk.): 'causing, responsible for' + **-logia** 'study of': the study of causes, in subjects from theology to physics, but especially in medicine, the aetiology of diseases, as in 'Severe acute respiratory syndrome coronavirus 2 (SARS-CoV-2) is the aetiological agent of COVID-19.'

ALIQUOT (L.) 'some, several, a few, not many'; typical context is the removal of a sample of a solution, i.e. an aliquot, for analysis, the remainder being subjected to further treatment, often depending on the analysis of the aliquot.

ALOPECIA: **alopekia** (Gk.) 'fox mange'. In humans, *alopecia* is the absence of hair from where it normally grows. The most common form of hair loss is *androgenetic alopecia* or male baldness. Another form is the recessively inherited *alopecia universalis*, where there is no body hair, eyelashes, or eyebrows, the apparent cause being absence of a transcription factor.

ANGLE: **angulus** (L.) 'corner, bend'; **quadrangulus** 'four-cornered'; *triangle, rectangle*, etc.; **gonia** (Gk.) 'corner': *trigonometry, polygon, octagon, diagonal*, and *orthogonal*, from **orthogonios** 'right-angled'.

ANHYDROUS: **anudros** (Gk.) from **an-** 'without' + **hudor** 'water'. Anhydrous means literally 'without water' and typically is applied to chemical reagents to indicate that they are pure or lack water of crystallization or are useful for reactions inhibited by water.

ANIMAL: **anima** (L.): 'air, breath, life'. The verb *animate* can mean 'bring to life', *animated* means 'lively', and *inanimate* means 'not alive'. *Animus* and *animosity* both indicate ill will or hostility; *unanimous* means everyone agrees.

ANOMALOUS: via late Latin **anomalus** from Greek **an-** 'not' + **omalos** 'even' as in *anomalous result*—an unexpected or difficult to interpret result.

ANOREXIA: **an-** 'not' + **orexis** (Gk.) 'appetite'. *Anorexia nervosa* is a mental disorder characterized by the lack of a desire to eat, in which *anorexigenic* neurons inhibit normal food appetite. A condition known as ORTHOREXIA consists of an unhealthy obsession about eating the 'right' food.

APPEAR: via OF, from **apparere** (L.), i.e. **ad** 'to' + **parere, -ui, -itum** 'to come into view' or 'become visible'. Similarly: *appearance, apparent, apparition*, and *transparent*, which describes materials such as clear glass or water which transmit light completely, cf. *translucent*, from **lucere** 'to shine', meaning *semitransparent*. *Parent* corresponds to the present participle of **parere**, meaning in this context 'bringing forth', as in *oviparous*; cf. **phanein** (Gk.) 'to appear'—see heading for TRYPTOPHAN under AMINO ACIDS.

ARCHA-: **arche** (Gk.) 'origins': *Archaea, archive, archaic, archaeology, archetype* (the original model or version).
ARCH- as in monarch: **archon** (Gk.): 'chief, ruler'; *archbishop, architect, archenemy, matriarch, patriarch, anarchy, archipelago* (literally 'chief sea', meaning for the Greeks the Aegean), *hierarchy*, from **hieros**: 'sacred'. 'Stem cells sit at the apex of a differentiation *hierarchy*.'

ARTIODACTYLA: **artios** (Gk.) 'even' + **daktulos** 'finger, toe'. The mammalian order of even-toed *ungulates* (hoofed animals **ungula** L. 'hoof') including *sheep, cattle, pigs, deer*, altogether some 200+ species; cf. **PERISSODACTYLA**: **perissos** 'uneven'. The odd-toed ungulates include *horses, tapirs, rhinoceroses*.

ASTRONOMY: **astron** (Gk.) 'star' + **nomos** 'distributing, arranging' from **nemein** 'to arrange'; cf. ASTROLOGY. AUTONOMOUS: **auto-** 'self or own' + **nomos** 'law'. ECONOMY: **oikos** 'house' + **nomos** where the sense is to manage or control. ECOLOGY has the same **oikos** root. TAXONOMY: **taxis**: 'order, rank' from **tassein** 'to classify' + **nomos**. From the **taxis** root: *tactic, chemotaxis*, and *ataxia* (loss of coordination of the extremities).

BAD: malus (L.): *malice, malign, malignant, malady, malnutrition*; see also *malaria* under MAL- in chapter 3; **kakos** (Gk.): *cacophony*, harsh discordant sounds; *cachexia*, body wasting due to chronic illness.

BUBO: from **boubon** (Gk.): groin. Inflamed abscesses in parts of the body containing lymph nodes, especially the groin and armpits, are a symptom of plague, i.e. infection by *Yersinia pestis*, hence referred to as *bubonic plague* and the abscesses as *bubos*. See *The Plague* by Albert Camus for more on bubos.

CALOMEL is the chemical mercurous chloride, Hg_2Cl_2. The name comes from **kalos** (Gk.) 'beautiful' + **melas** 'black'. Calomel is a white powder, but the name, coined by the alchemists, is because it turns black on reaction with ammonia. Calomel electrodes serve as reference electrodes, e.g. in pH meters.

CARE: cura (L.): 'care, attention': *cure, secure, curator, curious*.
accurate: from **accuratus** 'done with care' from **ad** + **cura**; curable vs. treatable—a chronic disease may be treatable but not curable.

CEREALS: named for Ceres, the Roman goddess of agriculture. The term *cereal* applies to any *monocotyledonous* plant, i.e. having a single seed leaf, whose seeds serve as food, such as wheat, barley, oats, rice.

CHEMOTAXIS: tassein (Gk.) 'to arrange' is the phenomenon whereby cells move towards higher concentrations of food or other chemical *attractants*. Chemotaxis is important during the early phases of *embryogenesis* as development at this stage is guided by *gradients* of *morphogenetic* molecules. Also from this root are *tactic* and *taxonomy*.

CHIMAERA or CHIMERA: originally, in Greek mythology, referred to a fire-breathing she-monster consisting of a lion's head, a goat's body, and a serpent's tail. Following the development of recombinant DNA techniques, DNA molecules containing sequences from different sources can also be described as *chimeric sequences*. More generally, a biological chimaera is defined as an individual, organ, or part with a diverse genetic constitution.

Cis (L.): 'on this side'; as in *cis-acting* factors, cf. *trans-acting* factors; also *cis-trans* isomerism at double bonds, see under FATTY ACIDS. A typical situation is a cell infected by a virus where the virus-encoded polymerase is *cis-acting* while the cell-coded immune response acting on the virus is *trans-acting*.

CLAUSTROPHOBIA: the irrational fear of confined spaces from **claudere, clausum** (L.): 'to shut or to close', which gave rise to the following family of verbs: *include, exclude, conclude, seclude, preclude, recluse*, also *clause*. A related root is **clavis** 'key'. The diminutive is *clavicle*, i.e. *collarbone*, allegedly having a similar shape to that of a key; also *autoclave* because it is self-locking, no key is required.

CONCENTRATE: chemical usage: to concentrate a substance, e.g. by removal of solvent, i.e. the opposite of *dilution*. The concentration of *heat-sensitive* biological material may involve removal of solvent by *freeze drying*, *ultrafiltration*, or *precipitation*. The product of the procedure may be referred to as the *concentrate*.

CONTAGIOUS from **contagio, -onis** (L.): 'touch, contact' vs. **INFECTIOUS**: Used in connection with disease, with a similar meaning to *infectious* (*inficere, infectum*, L.) except that *contagious* usually refers to the person or animal with the disease, contact with whom may lead to transmission of the disease. *Infectious* usually refers to the agent or organism that causes the disease. So the flu virus is infectious, while a person with the flu is contagious. That was the case before the arrival of COVID-19. Since then, these terms have often been used interchangeably.

CREATURE: **creator** (L.): founder: *create, creative, procreate, recreation*.

DEATH: **nex, necis** (L.): 'death'; *pernicious, necropolis, necroptosis, necrosis*; **thanatos** (Gk.): *euthanasia, thanatophoric* (death bearing).

DOXA (Gk.): 'opinion', from which we get *orthodox, heterodox,* and *paradox*. ORTHODOX: **ortho** 'straight or right' + **doxa**, hence a correct, received, or approved opinion. HETERODOX: **hetero** 'other'+ **doxa**, unorthodox, not conforming to received opinion. PARADOX: **para** 'distinct from' + **doxa**, used in relation to an unexpected result that does not fit with the current paradigm, the result of a line of reasoning that leads to a self-contradiction.

DRAWING: graphe (Gk.) 'drawing or writing': *graphite, graphic, biography, radiography, geography, autoradiograph.*

DRINK: pinein (Gk.) 'to drink'; **potos** *n.* 'drink'; **symposion** for the Greeks was a drinking party and has come across in English as *symposium*, which is often similar after the last talk. **Pinein** has been co-opted to form the term *macropinocytosis*, which is a regulated form of *endocytosis* whereby liquid is taken into cells in vesicles formed by invagination of the cell membrane. The related **potare** (L.) has given rise in English to *potable*, i.e. safe to drink. Latin has **bibere** 'to drink', hence *imbibe* and *bibulous*. Latin also has **ebrius**, *adj.*, as well as **intoxicare**. To be *inebriated* or *intoxicated* is slightly more respectable than being simply *drunk* (OE).

EGG: ovum (L.): *oval, ovary, ovoid, ovulate,* and *oviparous* (species that lay eggs, as distinct from those that are *viviparous*); **oon** (Gk.): *oogenesis, oocyte.*

EDEMA: oidema (Gk.) from **oidein** 'to swell': excess watery fluid in body tissues. Normal distribution depends on balance between fluid in tissues and in the blood. Edema, also spelt *oedema*, can result from heart, kidney disease or injury, anything that affects blood pressure, e.g. *pulmonary edema*. Treatment is with *diuretics*, i.e. drugs that increase urination, from **diourein** (Gk.) 'to urinate' from **dia-** 'through' + **ouron** 'urine'.

ELUCIDATE: from **elucidare** from **ex-** + **lucidus** 'lucid' from **lucere** 'to shine' from **lux, luc-** 'light', hence *lux, pellucid, translucent, lucid.*

EMPTY: vacuus (L.); *vacuum, vacuous, vacuole, evacuate.*

ENCYCLOPAEDIA: derived from **enkuklios paideia**, literally 'all-round education'. **Paideia** was used by Plato to mean culture, learning, and accomplishments.

EXIST or **TO BE: esse** (L. irregular verb), hence *absent*: from **absens, absent-**, present participle of **abesse**, from **ab-** 'away' + **esse** 'to be'.

Likewise, *interest* from **interesse**. Also, *essence* and *essential* from **essentia** 'the being or essence of a thing'.

FAR: ultra (L.): *far, extreme* as in *ultraviolet*; **ulterior**: *further*, also with the meaning 'intentionally concealed' as in *ulterior motive*; **ultimus**: *furthest* or *final*, as in *ultimate, ultimatum, antepenultimate*.

FEVER: febris (L.): *febrile, feverish*; **pyretos** (Gk.): *pyrexia, pyretic*, also **pyros** 'fire'; see **PYR-** entry in chapter 5 for compounds derived from this root.

GNOTOBIOTICS: gnotos (Gk.) 'known' + **bios** 'life', includes the study of both 'germ-free' animals and those whose microbial flora can be completely specified, cf. *axenic*.

GOOD: bonus (L.): **melior**: 'better', hence *ameliorate*, **optimus**: 'best'; *optimum, optimal, optimism*.

HEAT: calor, -is (L.): *calorie, calorimeter*; **therme** (Gk.): *thermal, thermometer, thermostat, thermodynamics*.

HYALINE: hyalos (Gk.) 'glass'. A type of semi-transparent cartilage is hyaline cartilage. According to Andrews (1947), the word was used by Egyptians to refer to a type of clear precious stone used in jewellery. When real glass was invented, it is said that it was Plato himself who named it **hyalos** after the precious stone.

IATROGENIC DISEASE: iatros (Gk.) 'physician': a disease caused by a doctor and hence *iatrogenic*. The doctor would prefer to call it an unforeseeable or an unavoidable side effect of the treatment. The following words from the same root could summarize three stages of life, not necessarily in this order: *paediatric, geriatric*, and *psychiatric*.

LAPSE from **labi, lapsum** (L.): 'to slip or slide'; *relapse, collapse, elapse, prolapse*.

LETTER: gramma, grammat- (Gk.) 'written character'; *epigram, monogram, diagram, grammar*.

MEMORY: memoria (L.) and **memor**: 'mindful'; *memorial, memorandum, memoir, commemorate*; **mnemon** (Gk.): 'mindful'; *mnemonic*, i.e. memory aid, *amnesia* and *amnesty* from **amnestia**: 'forgetfulness'. **Mnemosyne**, goddess of memory, was a Titaness, one of the immortals.

MIDDLE: medius (L.) hence *medium, median, mediaeval, mean, mediocre*. **Mesos** (Gk.): *mesoderm, mesentery, Mesozoic*.

MIME; 'to imitate or copy', from **mimeisthai** (Gk.) 'to copy'. *Mimesis* in biology is the imitation of behaviour of one species by another; *mimetic* from **mimetikos**, in pharmacology and biochemistry, describes a synthetic compound which has the same effect or function as a naturally occurring one; *biomimetic synthesis* involves approaches to the synthesis of compounds in the laboratory based on steps or strategies employed *in vivo*.

MIND: mens, ment- (L.): *mental, dementia*; **phren** (Gk.): *phrenology, schizophrenia*; **psyche** (Gk.): *psychosis, psychology, psychotropic* (as in drugs).

MIXTURE: from **miscere, mixtus** (L.): *miscible, miscellaneous, promiscuous*.

MOLECULE: diminutive of **moles** (L.) 'mass'. The term was first used by Gassendi (1592–1655) in a treatise on the theory of atoms. It was adopted by the so-called corpuscular or particulate philosophers, adherents of the atom theory who took it to mean the smallest unit of a substance capable of a separate existence. The word *demolish* also comes from the **moles** root.

NARROW: stenos (Gk.): *stenosis*—narrowing or constriction of a passage or vessel of the body, reversible by insertion of a *stent*.

NEAR: prope (L.), **proprius** 'nearer', **proximus** 'nearest', hence *propinquity, proximity, approximate*, and *proximal*, meaning 'situated towards the point of attachment' for a structure such as a limb. The opposite of proximal is *distal*. From **proprius**, we get *approach*, via OF from **appropriare** and *appropriate*, which has two different meanings: if the second *a* is pronounced short, the meaning is 'fitting' or 'suitable for a particular purpose'; if the *a* is long the meaning is 'to take possession of', which probably derives from a second meaning of **proprius** 'one's own' as in *property* and *proprioception*. See entry for **AESTHETE**-, chapter 5.

NIGHT: nox, noct- (L.): *nocturnal, equinox,* cf. *diurnal.*

NOSE: nasus (L.): *nasal;* **rhis, rhinos** (Gk.): *rhinitis, rhinoplasty, rhinorrhoea.*

PARAMETER. Derived from the Greek, where it meant to measure one thing by another or to compare. Multiple meanings in English depending on subject, but more generally, a feature or characteristic that can be measured.

PARAFFIN: waxy solid or oil, mixture of saturated hydrocarbons; from **parum** 'little'+ **affinis** 'partaking in', named on account of its low reactivity.

PATIENT: patior, pati, passus (L.) 'to endure, suffer, bear, undergo'. It is easy to see how both meanings of *patient* arose. The present participle of **patior** is *patiens;* the adverb *patiently* is **patienter**. The past participle **passus** is the root for *passive,* as in *passive immunity,* the short-term immunity resulting from injection of exogenous antibodies.

PUBLISH: via OF from **publicare** (L.) 'to make public', but *publication* direct from Latin. The word *publish* is liable to bring to the mind of a scientist the phrase 'publish or perish', and the common final syllable suggests correctly that *perish* also spent time in France and derives ultimately from **perire** 'pass away', from **per-** 'through, completely' + **ire** 'to go'. The past participle of **ire** is **itum**, which provides the stem for *ambition, coition, exit, initial, obituary, sedition,* and *transition;* similarly, from the corresponding present participles: *ambient* and *transient.*

PULSE from **pellere, pulsi pulsum** (L.) 'to drive or to beat', defined by the *OED* as the rhythmically recurrent dilation of the arteries as the blood is *propelled* along them by the contractions of the heart. Also from this root are *repulse, compulsion, impulse.* **Sphygmos** (Gk.): 'pulse' from **sphyxein** 'to beat or throb', hence *sphygmomanometer* and *sphygmograph* to measure and record pulse and *asphyxiate* from **asphyxia**: 'the stopping of pulse and respiration'. See also SPHINGOSINE under **MEMBRANES.**

REMEDY: from **remedium** (L.), from **re** 'back' + **medeor** 'to heal'. The same stem gives *medicine, medical,* etc.

SCHEDULE: from late Latin **schedula** 'slip of paper', diminutive of **scheda**; from Greek **skhede** 'papyrus or sheet of paper'.

SEPARATE: **cernere, cretus** (L.) and so **secernere, secretus**, hence *secrete*; **discernere**: *discern, discrete*, meaning individual or distinct as in 'it gave a discrete band on a polyacrylamide gel'; *discreet* means tactful or cautious or 'to act with discretion'; also **excernere**: *to excrete*; **concernere**: *concern* but *concrete* derives appropriately from **concrescere**, 'to become harder'. **Krinein** (Gk.): 'to separate' and which has given rise a group of words connected with secretion: *angiocrine, autocrine, endocrine*, etc. See under **SECRETION MECHANISMS.**

SMALL: **minor** (L.); **minus**: 'less'; **minimus**: 'least'; *minority, minimal, diminish, miniscule*; **micros** (Gk.): *microbe, micron, microtome, microscope*.

SMELL: **ozo** (Gk.) 'smell'; **osme** (Gk.) 'odor, scent'; *ozone, osmium, anosmia* (loss of sense of smell); **odoratus** (L.): 'odor'. OLFACTION: the act of smelling or the sense of smell.

SOUND: **tonos** (Gk.): 'tone', from **tenein** 'to stretch', *monotone, baritone, peritoneum, tonic*; **phone** (Gk.): *phonetics, euphony, telephone, cacophony, symphony*. **Sonus** (L.): 'sonic' as in sonic boom, *sonicate, resonate, sonorous, sonar*.

STAND: from **stare, statum** (L.): 'to stand': *station, stationary, interstate, stationary, statistics, constant, distant, extant, instant, substantial, assistance, circumstance*; **stasis** (Gk.): *ecstasy, haemostat, homeostasis, prostate, thermostat*.

SWIFT: **celer** (L.): *accelerate*; **tachys** (Gk.): *tachometer, tachycardia* (abnormally fast heart rate).

TERATOGEN: from **teras, terat-** (Gk.) 'monster': agent causing embryonic malformation; *teratogenesis*

TETRAHEDRAL: **tetraedros** (Gk.) 'four-sided', the adjective corresponding to *tetrahedron*, one of the five platonic solids along with the cube, octahedron, dodecahedron, and icosahedron, containing four, six, eight, twelve, and twenty faces respectively. They are called platonic solids because Plato speculated that these were the elements of which the universe was constructed. For methane, CH_4, the four hydrogens are located at the vertices of a tetrahedron.

THERIA: **theria** (Gk.) 'wild animals'. The group of mammals that includes the marsupials and placentals. EUTHERIA: **eu-** 'true' + **theria**. The group of mammals comprised of all the placentals. METATHERIA: **meta-** 'expressing change or difference' + **theria**. The group of mammals consisting of the marsupials. PROTOTHERIA: **proto** 'first, original' + **theria**. The group of mammals comprised of the monotremes, the platypus and echidnas and their extinct relatives. The monotremes, **mono-** 'single' + **trema** 'hole', are egg-laying mammals and have a common opening for their urogenital and digestive systems.

THESIS: from **thetos** 'placed' and **tithenai** 'to put or place', hence *antithesis, hypothesis, parenthesis, prosthesis.*

VALENCY: **valere** (L.) 'to be strong'. Valency is the combining power of an element, described as a foundation doctrine of modern chemistry. Hydrogen has a valency of 1, magnesium 2, etc.

ZYGOTE: **zygon** (Gk.): 'yoke', from **zugoun** 'to yoke two animals together with a wooden harness joined to a wagon or plough', *homozygote, heterozygote.*

CHAPTER 5

Absorb to *Zoonoses* –
Terminology in Context

HEADINGS

ABSORB, ADSORB, and
ABSORBANCE
ACETIC ACID
ADIPOSE TISSUE
ADRENAL GLANDS
AESTHET-
AGGREGATE
ALIPHATIC and AROMATIC
AMINO ACIDS
AMMONIA
ANALGESIA
ANTHROPO-
ANTIBIOTIC
ANTIOXIDANTS
APICOMPLEXA
ARCHAEA
ASCORBIC ACID
AUTOIMMUNE DISEASES
BACILLUS
BILE
-BLAST and BLASTO-
BLOOD
CADAVER
CANCER
CARBOHYDRATES
CARBON
CARTILAGE
CENTRIFUGE

CEPHALO-
CHELATION
CHEMISTRY
CHIRALITY
CHOLESTEROL
CHROMATOGRAPHY
CHRONIC
CHYME
CIRCADIAN RHYTHM
CIRCLE
COLLOID
COLOURS
COMPOSITION and
COMPOUND
CRYO-
CYCLE
CYTO-
DERMA
DIMINUTIVES
DROMOS
DYSLEXIA
ELECTRO-
ELEMENT
ELUTE
ENDOPLASMIC RETICULUM
ENTERO-
ENZYMES
ETHYL-

FATTY ACIDS
FLOW
FLUX, FLUORINE, and
FLUORESCENCE
FORMULA
FRAGMENT
FREE ENERGY
GENOMES
GENOMICS
63. GLUCAGON and
GLYCOGEN
GLYCOLYSIS and
GLUCONEOGENESIS
GLYCOPROTEINS
HELIUM
HISTO-
HOMOLOGY
HYDROGEN
HYPERSENSITIVITY
IMMUNITY
INFLAMMATION
INTEGER
IRON
ISOPRENE DERIVATIVES
KARYO-
KINASE
LIGATE
LIPIDS
LIPOSOMES
-LOGY, -LOGICAL, -LOGIST
LYSIS
MEMBRANES
MENTOR
METABOLISM
MEASURE
MILK
MITOCHONDRIA
MITOSIS and the CELL CYCLE

MONOCLONAL
ANTIBODIES
MORPHO-
MUCO-
MUSCLE
MUTATION
MYCO-
MYELIN
NAME
NATIVE
NOCEBO
NUCLEIC ACIDS and GENE
EXPRESSION
NUMBERS and UNITS
OESTRUS
-OME
OPSINS
ORGAN, ORGANISM,
ORGANIZE, ORGANELLE,
and ORGANOIDS
OSTEO-
OXYGEN
PAEDIATRICS
PATHO-
PEPTIDOGLYCAN
PHAGE, -PHAGY
PHARMACO-
PHEROMONE
PHORESY
PHOSPHORUS
PHOTOSYNTHESIS
PLOIDY
PNEUMO-
PROTEINS
PYR, PYRO-
RECEPTORS
SALT

SCIENTIFIC KNOWLEDGE:
THEORY and PRACTICE
SCOPE
SECRETION MECHANISMS
SELENIUM
SENESCENCE
SLEEP
SEQUENCE
SODIUM and POTASSIUM
SOLVENT
SOMATO-
SPEECH
SPORE
STEM CELLS
STEP
STOCHASTIC
SULPHUR
SYMBIOSIS
TERMINOLOGY
TISSUE
TOMOS
TOPO-
TROPH-
TROPISMS, TROPES
VIRUS
VITALISM
VITAMINS
WORDS of BECOMING
XENO-
XERO-
XYLO-
YEAST
ZOONOSES

ABSORB, ADSORB, ABSORBANCE: the ability of some materials to absorb liquid is familiar enough. The word comes from Latin: **absorbere**: **ab** 'from' + **sorbere** 'to suck in'. From the past participle **absorptum** comes the English noun *absorption*. The ability of organisms to absorb raw materials from the environment, the products of food digestion in the case of animals, is central to life. Molecules can cross membranes either by *active transport*, which uses the energy of ATP hydrolysis to pump molecules against a concentration gradient, or by *facilitated diffusion*. In both cases, specific membrane-bound *transporter* proteins are involved. In the bacterium *Escherichia coli*, the genome encodes about 4,000 proteins, of which 160 are transporters. In designing drugs, their ability to be absorbed across membranes needs to be considered.

Some common chemical reagents, such as calcium chloride, absorb moisture from the air and are said to be *hygroscopic*, from the stems **hygro-** (Gk.), meaning wet, and **skopein**: 'to look at or detect', while a *hygrometer* is an instrument for measuring humidity of the air or a gas. Another word with a similar meaning is *deliquescent*, from **deliquescere** (L.) 'to dissolve'. The difference between *hygroscopic* and *deliquescent* is one of degree—a salt that is deliquescent will eventually dissolve when exposed to the air. Thus, $CaCl_2.2H_2O$, the *dihydrate*, is described in the Merck Index as hygroscopic and the *hexahydrate* as deliquescent. These types of compounds find use as drying agents or *desiccants*, from the Latin **siccus** 'dry', **siccare** 'to dry', and **desiccare** 'to dry completely', or perhaps by using a *desiccator*.

ADSORB was derived from *absorb* in the nineteenth century to describe the situation where molecules or materials bind to the surface of a solid, the *ad-* indicating *adhesion*. *Chromatography* separates components of mixtures by differential adsorption to a solid support, e.g. paper, in the presence of a particular solvent. The Haber-Bosch process for the synthesis of **AMMONIA** depends on the binding of hydrogen and nitrogen to a solid catalyst at high temperature and pressure. *Activated charcoal* functions as an *adsorbent* in applications such as *decolourizing* and *decontamination*. Its effectiveness as an adsorbent comes from being extremely finely divided so that 1 gram can have a total surface area of up to 500 m². *Desorb* is the opposite of adsorb.

ABSORBANCE: the interaction of electromagnetic radiation with matter often results in the selective absorption of particular wavelengths. When this happens with visible light, we perceive colour. Wavelengths of light outside our visible spectrum cannot be distinguished by humans, although other wavelengths are easily perceived by bees in the UV range, or cats in the infra-red range. A solution of copper sulphate appears blue because it absorbs red light. The basis for this is that light of a particular wavelength is absorbed when its energy matches that required to shift electrons of the metal ion to a higher energy level. These effects depend strongly on the particular electronic environment: other compounds of copper, particularly cuprous compounds, exhibit a range of colours. Chromium (named from the Greek **khroma** 'colour'), manganese, iron, cobalt, nickel, and copper, all from the same row in the periodic table, form characteristic coloured compounds: iron is red, cobalt is blue, nickel is green, etc.

ACETIC ACID: **acetum** was Latin for vinegar, or sour wine, while **acidus** was an adjective meaning 'sour or tart'. Another word, **acer-**, **acri-**, meaning 'to be sour', was probably the original root for both **acetum** and **acidus** as well as for **acescere** 'to become sour' and **acerbus**, also meaning 'sharp, sour, or bitter'. The formula for acetic acid is CH_3COOH. Pure acetic acid is referred to as glacial acetic, **glacies** (L.) 'ice', as in glacier, because it forms ice-like crystals when cooled. The ionized form is acetate, formula CH_3COO^-, and is the basic two-carbon compound of intermediary metabolism. Acetate is usually encountered as the activated form in thioester linkage to Coenzyme A, i.e. acetyl CoA. In Rome, an **acetabulum** was a small cup for serving vinegar and other condiments, while in modern anatomy, it is the cup-shaped cavity in the hip bone to which the head of the femur fits.

ACIDOSIS refers to increased acidity, i.e. hydrogen ion (H^+) concentration, in the blood and other body fluids. Acidosis is defined as a pH of arterial blood below pH 7.35. Gastric acid consists mainly of hydrochloric acid with a pH of 1–2. ACRID also derives from **acer-**, **acri-**, which is the root for *acrylic acid* (2-propenoic acid, $CH_2=CH-COOH$) and derivatives of the amide, *acrylamide*, are widely used to make electrophoresis gels. *Acridine* is a *polycyclic* compound which, because of its ability to intercalate between DNA base pairs, can cause *frameshift mutations*. Acridine and some of its derivatives used to be marketed as *antiseptics* but are now discontinued because of their mutagenic activity. Related words: *acerbic, acrimonious*.

ADIPOSE TISSUE: adeps, adipis (L.) 'fat, grease, lard'. Adipose tissue is a type of *connective tissue* where the main function is the storage of energy in the form of *triglycerides*. It consists of fat-filled cells termed *adipocytes*. The main form of adipose tissue is white in colour (WAT). A second form is *brown adipose tissue* (BAT), where the function is *thermoregulation*. The brown colour is due to the presence of *mitochondria*, but here electron transport is *uncoupled* from ATP synthesis and the energy from fat oxidation is obtained as heat. In mammals, BAT is only found around the neck and *thorax*, and its occurrence in humans is restricted to infants. This may help the survival of *neonates* exposed to cold. It may also be used to rewarm the bodies of animals that *hibernate*. A second type of brown fat, known as *beige* or *inducible BAT* (iBAT), has recently been discovered within white fat deposits. The possibility of treating *obesity* by stimulating iBAT synthesis at the expense of WAT synthesis has been suggested. You would have a choice: either fat or hot.

Both fat synthesis (*lipogenesis*) and fat breakdown (*lipolysis*) are closely regulated. The initial step in synthesis is carried out by *acetyl CoA carboxylase*, which converts acetyl CoA to *malonyl CoA*.* This is the *rate-limiting step*, so by controlling the rate of this reaction, the overall rate of fat synthesis is determined. This enzyme is stimulated by *insulin* and *citrate*, indicators of energy abundance, and inhibited by *adrenalin* and *glucagon*. The remaining steps in *fatty acid* synthesis are carried out by *fatty acid synthase*, which, in mammals, consists of multiple enzyme activities within a single very large polypeptide chain. The main product is *palmitic acid*. The rate-limiting step for *energy mobilization* is carried out by the enzyme *hormone-sensitive lipase* (HSL), which removes a single fatty acid from triglycerides. The lipase is activated by *catecholamines* and *adrenocorticotropic hormone*, ACTH, but is inhibited by *insulin*. In obesity, **obesitas** (L.) 'fatness', adipocytes increase in size (*hypertrophy*), and to a lesser extent, they proliferate, i.e. *hyperplasia*.

*Malonic acid: obtained by oxidation of *malic* acid from **malus** (L.) 'apple'.

ADRENAL GLANDS: these are *endocrine glands* situated above the kidneys, **ren, renes** (L.) 'kidney', so **ad + renalis**, 'of or near the kidneys'. The adrenal hormones ADRENALIN and NORADRENALIN are produced in the *medulla* of the adrenal glands, the term *medulla* referring to the internal portion of an organ. These hormones are referred to by American authors as *epinephrine* and *norepinephrine*, **epi-** 'above' + **nephros** (Gk.) 'kidney'. The *nor-* prefix, short for *normal*, is used by chemists to discriminate between two structures that differ by a single peripheral carbon atom. Adrenalin and noradrenalin act as both hormones and neurotransmitters. They are mediators of the complex 'flight or fight' response, which involves their binding to several receptors, the *adrenergic receptors*, which are themselves distributed in different tissues.

The outer part of the adrenal gland is the *cortex*, a term in Latin for the bark of a tree. The adrenal cortex is the site of synthesis of the CORTICOSTEROIDS: these consist of two distinct types: *glucocorticoids* namely *cortisol* and *corticosterone* on the one hand and *mineralocorticoids* on the other. Glucocorticoids have wide-ranging effects on metabolism, the cardiovascular system, and the immune system. They function by binding to specific receptors in the cytoplasm, which then enter the nucleus to act as *transcription factors* controlling gene expression. See entry for NUCLEAR RECEPTORS under **RECEPTORS**. They stimulate **GLUCONEOGENESIS**, lipolysis in **ADIPOSE TISSUE** and serve as *anti-inflammatories*, which make them, as well as their *synthetic analogues*, useful for the treatment of asthma and other allergies. The main mineralocorticoid is *aldosterone*, which acts on the kidney to regulate the retention of sodium, potassium, and water and thereby blood pressure and blood volume. The biosynthetic pathway from **CHOLESTEROL** to the corticosteroids begins with truncation of the cholesterol side chain to just two carbons, to yield C21 *pregnenolone*, the precursor of *progesterone*, which in turn gives rise to the corticosteroids. For more on corticosteroids see also https://en.wikipedia.org/wiki/glucocortioid and https://en.wikipedia.org/wiki/mineralocorticoid

CATECHOLAMINES. Catechol is 1,2-dihydroxybenzene. A catecholamine consists of catechol with a side chain containing an amino group. The three important catecholamines are dopamine, noradrenalin, and adrenalin. For their structures, see https://en.wikipedia.org/wiki/ Catecholamine.

AESTHET-: aisthetikos (Gk.) 'of sense perception', **aesthesis**: 'sensation' AESTHETICS: to do with the perception and appreciation of beauty. ANAESTHESIA decreases sensory perception, including the perception of pain and, with general anaesthesia, loss of consciousness. Humphry Davy (1778–1829) in 1799 suggested the use of nitrous oxide as an anaesthetic, but it was used only recreationally until 1845, when it was first used in dentistry. It continued to be used in dentistry for well over one hundred years. Ether was first used as an anaesthetic in an operation in 1846, and chloroform was also introduced soon after.[11] Both ether and chloroform have been replaced by other volatile anaesthetics. CHEMAESTHESIA is a recent coinage for the situation where a chemical mimics a sensory stimulus, an example being the action of *capsaicin*, from chilli pepper, which induces a hot sensation when bound to the TRPV sensory receptor, which normally responds to an increase in temperature by opening a calcium channel. See entry for **RECEPTORS**.

KINAESTHESIA: **kinein** (Gk.) 'to move' describes awareness of the position and movement of the body by means of sensory organs (*proprioceptors*) in the muscles and joints. The term *kinaesthesia* seems likely to be replaced by PROPRIOCEPTION: **proprius** 'one's own' + **-ception** as in receptor or perception. PHONAESTHESIA: **phone** (Gk.) 'sound or voice' refers to feelings associated with particular sounds, usually the sounds associated with the pronunciation of particular words. Burnside[12] gives the example of words where the dominant syllable, or *phoneme* as the linguists say, is *-unk*. Words such as *drunk, funk, gunk, hunk, junk, punk, skunk, slunk*, which comprise most of the commonly used English words that end in *-unk,* all have a somewhat negative connotation, probably an example of low-level phonaesthesia. SYNAESTHESIA: **sun-** (Gk.) 'together' is a rare familial condition in which stimulation of one sense, such as listening to music, is accompanied by perception of another sense such as taste or colour. Thus, synaesthetic tastes elicited by words are said to be experienced by *lexical-gustatory* synaesthetes (**lexis**: Gk. 'word'; **gustare** L. 'to taste'). Synaesthesia has been described as 'a fusion of the senses'.

AGGREGATE: aggregare (L.) 'to herd together' from **grex, greg-** 'flock, herd or swarm'. In English and in the context of protein chemistry, the verb 'to aggregate' can indicate the herding together of protein molecules, or as a noun, it can mean the resulting complex, the aggregate. Although many, if not most, proteins perform their functions in *complexes* containing multiple components, the terms *aggregate* and *aggregation* usually imply that protein *denaturation* has occurred. Another example is *sickle cell haemoglobin* where, as the result of a *mutation* causing a single amino acid substitution, haemoglobin molecules form large aggregates that deform the red blood cell, causing it to assume a sickle shape. A variety of neurodegenerative human diseases, also known as *amyloid* diseases, such as Alzheimer's disease, are characterized by protein aggregation.

The **grex, greg-** root also gave rise to *egregious*. Originally, *egregious* meant something or someone that stood out because of exceptionally good qualities, but *egregious* now describes a standout for exactly the opposite reason. Another **greg-** word is *gregarious*, usually an adjective applied to people but also to animals that *congregate* such as sheep (flocks), cattle (herds), fish (schools), and bees (swarms). Not coincidentally, gregarious animals are those that man has found easiest to exploit. Also *segregate*: **se-** meaning 'apart'.

ALIPHATIC: aleiphar, aleiphat- (Gk.) 'fat', denotes carbon compounds such as the series methane, ethane, propane, etc. as well as those containing one or more double bonds such as ethylene and acetylene. *Fatty acids* consist of a terminal carboxyl group together with an aliphatic tail or side chain. The historical justification for the term *fatty acid* is that the first members of this class to be studied were obtained from fats, i.e. *triglycerides*. Biologists still use the term, but chemists prefer the term *aliphatic acid*. The chemists then decided that non-aromatic cyclic compounds like cyclohexane should be called *alicyclic* compounds because they combined aliphatic and cyclic properties while compounds like fatty acids and straight-chain hydrocarbons should be called *acyclics*.

AROMATIC: aromatikos (Gk.) from **aroma** 'spice'. Aromatic hydrocarbons can be represented by the presence of a ring structure consisting of alternating double bonds as in representations of *benzene*, C_6H_6. This arrangement indicates *electron delocalization* that confers chemical properties distinct from those of aliphatic hydrocarbons. The odour of benzene itself can hardly be described as aromatic, but it was realized early that the active components of aromatic natural oils were benzene derivatives. It was the German chemist Kolbe who in 1860 distinguished aromatic acids from fatty acids, citing the smell of the *aldehydes* of the former to justify this name. In 1865, Kekule proposed the hexagonal ring structure for benzene.

AMINO ACIDS: the α-*amino acids* are the ones we are interested in here, where both the amino group and the carboxyl group are attached to the carbon atom designated as the α-carbon atom. These are the two groups that react to form a *peptide bond* during *protein synthesis*. Subsequent *side-chain* carbon atoms are β-, γ-, δ- carbons, while lysine, which has a total of six carbons, including the carboxyl group, has the side-chain amino group attached to the ε-carbon atom. Numerous other amino acids occur naturally but not as constituents of proteins. Because amino acids contain both *basic* groups, i.e. amino groups that can accept protons and *acidic* groups that can donate protons, they are said to be *amphoteric*, from the Greek **amphoteros** meaning 'both'. Another term is **zwitterion**, from the German, meaning hybrid. Because the pKa values of amino groups are in the range 8–10, and those for carboxyl groups 3–4, it means that at neutral or physiological pH, amino groups will be protonated, $-NH_3^+$, and carboxyl groups will be dissociated $-COO^-$. It follows therefore that the simple formula $NH_2-CH(R)-COOH$ does not represent a species that will be present at any pH. For the structures of the twenty-one amino acids that occur in proteins, go to https://en.wikipedia.org/wiki/Amino-acid.

The net charge on proteins depends on the relative abundance of the acidic and basic amino acids, and the pH at which the net charge on a protein is zero is referred to as its *isoelectric point*. Histones are examples of proteins with a high proportion of lysine and arginine residues and are therefore strongly positively charged at neutral pH, in line with their function, which is to associate with negatively charged DNA in the *nucleosome*. Apart from their role as constituents of proteins, several amino acids have roles in nitrogen and carbohydrate metabolism and as precursors for other amino acids.

GLYCINE: **glukus** (Gk.) 'sweet'. This is the simplest amino acid and has no *asymmetric centre*. All other amino acids have four *different* substituents at the α-*carbon atom*. Depending on the configuration at the α-carbon atom, amino acids are either D- or L-. See the entry for *threonine* below, also the entry for **CHIRALITY**. All amino acids in proteins have the L- configuration. Glycine, with just a hydrogen side chain, is able to occupy positions where the peptide chain turns sharply and *steric hindrance* between side chains would otherwise occur. This is also the reason that glycine occurs at every third residue in the fibrous protein *collagen*, which consists of three polypeptide chains wound tightly together.

PROLINE: contraction of *pyrrolidine-2-carboxylic acid*. Proline, although usually referred to as one of the *amino* acids normally found in proteins, is really an *imino* acid with its side chain joined to the α-amino group to form a *secondary amine*. Proline tends not to occur within α-*helices* or β-*sheet* structures of proteins. Both these structures are stabilized by *hydrogen bonds* involving the O and the H of the -CO-NH- *peptide bond* and so would be destabilized by the presence of proline residues lacking an H in this position. Proline is a major component of collagen, which is the most abundant protein in mammals.

Amino acids with aliphatic side chains

ALANINE: the *al-* because it was first synthesized from an aldehyde, the *-an-* inserted for 'reasons of euphony'. The corresponding α-keto acid is *pyruvate*.

VALINE: Another name for valine is α-amino isovaleric acid, which in turn takes its name from the plant valerian.

LEUCINE and ISOLEUCINE are isomers, but it was nearly a century before they were distinguished (see table 8), both named for their white (**leuko-** Gk.) crystals, hardly a distinguishing characteristic. Note that isoleucine has two asymmetric carbons, so there are four possible isoleucines, only one of which is found in proteins. *Hydrophobic side chains* as in these amino acids are usually located in the interior of folded proteins and out of contact with water.

Basic amino acids

HISTIDINE: from the Greek root **histos** 'tissue', cf. *histology* and *histone*. Histidine occurs in the *active sites* of numerous enzymes. The *imidazole* side chain is partially *aromatic* which, along with a side-chain pKa near neutral, helps to explain the frequent occurrence of histidine residues at enzyme active sites.

ARGININE: from **argos** (Gk.) 'shiny white', referring presumably to its crystals. The side chain contains a positively charged *guanidinium* group so that arginine, like the other basic amino acids, can form a *salt linkage* with an acidic amino acid. Arginine is also an intermediate in the urea cycle.

LYSINE: from 'lysis', in turn from **luein** (Gk.) 'to loosen', presumably alluding to its original isolation from an *acid hydrolysate* of protein. Because of the presence of the ε-*amino group*, lysine residues in proteins are subject to a variety of *post-translational modification* reactions. Specific lysine residues in the *histones* are subject to *acetylation* and *methylation* as part of *epigenetic* control of gene expression.

The acidic amino acids and their amide derivatives

GLUTAMIC ACID: the name derives from the material from which it was first isolated: the 'glut' is from *gluten*, a family of proteins found in cereal grains but earlier on was a non-specific term for materials including gums and glues—**glus, glut**- (L.) 'glue'. The corresponding *keto acid* is the citric acid cycle intermediate α-*ketoglutarate*. In only a handful of proteins, some glutamyl residues are post-translationally modified to form γ-*carboxy-glutamyl* (three-letter abbreviation Gla). See entries for BLOOD CLOTTING and **OSTEO**-.

GLUTAMINE: from *glutamic amide*, i.e. the γ-*carboxamide* of glutamic acid. ASPARTIC ACID. The keto acid corresponding to aspartic acid is the citric acid cycle intermediate *oxaloacetate*. The side-chain carboxyls of aspartic and glutamic acids are completely ionized at neutral pH.

ASPARAGINE: from the Greek for asparagus, **asparagos**. According to table 8 below, it was characterized well before aspartic acid, of which it is the amide, was itself recognized. The amide groups of both glutamine and asparagine are lost during the standard conditions of acid hydrolysis used to determine the amino acid composition of proteins.

Amino acids with aliphatic side chains containing hydroxyl groups

SERINE: **serica** (L.) 'silk'. Serine was first isolated from hydrolysed silk. The Romans referred to the people of eastern Asia as the Seres, i.e. the modern Chinese, whose silken fabrics were prized by the Romans. The cultivation of silkworms is known as *sericulture*. Both serine and threonine can be phosphorylated on their side chain OH groups, or linked to carbohydrate residues. Carbohydrate-containing proteins are known as **GLYCOPROTEINS**. Linkage of the carbohydrate (*glycan*) is either via the amide group of asparagine, i.e. N-linked, or via the hydroxyl group of serine or threonine, i.e. O-linked.

THREONINE was originally isolated from casein hydrolysates in 1935 as a factor essential for the growth of rats. It was shown to have the same stereochemistry about the β-carbon as the sugar D-*threose* and was named accordingly. (In spite of this, threonine as it occurs in proteins is designated L-threonine. This is because D-sugars are so named when they have the same configuration about the highest asymmetric carbon atom while amino acids are designated L or D depending upon their configuration about the lowest asymmetric carbon atom, i.e. the α-carbon).

The sulphur-containing amino acids

CYSTEINE: **kustis** 'bladder'. Cystine, the oxidation product of cysteine, was first isolated from urinary calculi. *Disulphide bonds*, -S-S-, in proteins result when the -SH (*thiol*) groups of two cysteine residues react. Cysteine is a component of the tripeptide antioxidant *glutathione*. See entry for **ANTIOXIDANTS**.

METHIONINE: Its name derives from the presence of the terminal methyl group and the sulphur (**thio**) atom. Formula: $CH_3SCH_2CH_2CH(NH_2)$ COOH. An enzyme-catalysed reaction between ATP and methionine forms S-adenosylmethionine (SAM), which serves as a methyl donor in many methyl transferase reactions. The positive charge on the sulphur atom of SAM activates the methyl group. During protein synthesis on the ribosome, the first codon to be translated, i.e. the *initiation codon*, is AUG, the codon for methionine. This means methionine is the first amino acid to be incorporated into the nascent peptide chain (*N-formylmethionine* in bacteria). For most proteins, this terminal methionine is subsequently removed by a methionine *aminopeptidase*. AUG also codes for internal, i.e. downstream, methionines.

The (other) aromatic amino acids

The aromatic amino acids are phenylalanine, tyrosine, and tryptophan, and these can be regarded as derivatives of alanine, in which a hydrogen of the β-methyl group is replaced by the aromatic groups corresponding to *benzene*, *phenol*, and *indole* respectively. One might have expected the name for phenylalanine to be benzylalanine, and the reason that it is not has to do with how terminology developed in the early days of organic chemistry.[13]

PHENYLALANINE, being an essential amino acid, has to be supplied in the diet. Normally about 75 per cent of phenylalanine obtained in the diet is converted into tyrosine, the remainder incorporated into protein. Absence of the enzyme that converts phenylalanine to tyrosine results in accumulation of phenylpyruvate and the disease condition *phenylketonuria*. Untreated, this can lead to mental retardation. Therapy consists of a low-protein diet that is just sufficient to meet the requirements for phenylalanine and tyrosine.

TYROSINE: **turos** (Gk.) 'cheese', first isolated from casein, the major protein complex in milk and cheese. Tyrosine is often found in protein pockets or cavities that are involved in binding ligands, also antibody binding sites. Its ability to form both hydrophobic and hydrogen bond interactions as well as its size and conformational rigidity means it is well suited to roles involving molecular recognition. Tyrosine is the precursor for synthesis of the *catecholamines*. See entry for **CORTICOSTEROIDS**.

TRYPTOPHAN: trypsin + **phainein** 'to appear', the **phan** root attesting to the fact that tryptophan was first seen in *tryptic* digests of protein. The **phainein** root has given rise to several other English words, including *phenyl-* as in *phenylalanine*, *phenotype*, *diaphane* (i.e. something transparent, hence *diaphanous*), *epiphany*, *phenomenon*, and *cellophane*, the transparent polymer derived from treatment of cellulose with sodium hydroxide and carbon disulphide. The broad-spectrum herbicide *glyphosate*, marketed as Roundup, acts by inhibiting an enzyme on the pathway to *chorismate*, which is a precursor of all three aromatic amino acids.

SELENOCYSTEINE: the twenty-first amino acid occurring naturally in mammalian proteins. Selenium occurs in group VI under sulphur in the periodic table and selenocysteine (Sec) is the selenium analogue of cysteine in which the thiol (-SH) of cysteine is replaced by *selenol* (-SeH). It is encoded by UGA, which normally is a stop codon but codes for Sec when the adjacent mRNA sequence contains a specific stem-loop structure. Some twenty-five human proteins are known to contain selenocysteine. Several of these are *glutathione peroxidases*, which are crucial for removal of reactive oxygen species. See also entry for **SELENIUM**.

Essential amino acids for humans are those which have to be supplied in the diet. These are histidine, isoleucine, leucine, lysine, methionine, phenylalanine, threonine, tryptophan, and valine. *In vivo* degradation of most amino acids yields pyruvate or an intermediate of the citric acid cycle, and these are therefore said to be *glucogenic*. Others produce acetyl CoA and acetoacetate and are *ketogenic* or are both glucogenic and ketogenic.

In table 8 below, codons are the nucleotide triplets in messenger RNA that specify particular amino acids. See also the genetic code in table 9.

TABLE 8: THE TWENTY-ONE AMINO ACIDS

NAME	3-LETTER ABBREV.	1-LETTER SYMBOL	CODONS*	DISCOVERED
Alanine	Ala	A	GCX**	1850
Arginine	Arg	R	CGX, AGA, AGG	1886
Asparagine	Asn	N	AAU, AAC	1806
Aspartic acid	Asp	D	GAU, GAC	1868
Cysteine	Cys	C	UGU, UGC	1899
Glutamic acid	Glu	E	GAA, GAG	1866
Glutamine	Gln	Q	CAA, CAG	1883
Glycine	Gly	G	GGX	1820
Histidine	His	H	CAU, CAC	1896
Isoleucine	Ile	I	AUU, AUC, AUA	1904
Leucine	Leu	L	CUX, UUA, UUG	1819
Lysine	Lys	K	AAA, AAG	1889
Methionine	Met	M	AUG	1922
Phenylalanine	Phe	F	UUU, UUC	1879
Proline	Pro	P	CCX	1900
Serine	Ser	S	UCX, AGU, AGC	1865
Threonine	Thr	T	ACX	1935
Tryptophan	Trp	W	UGG	1901
Tyrosine	Tyr	Y	UAU, UAC	1846
Valine	Val	V	GUX	1901
Selenocysteine	Sec	U	UAG	1974

*See also Genetic Code Table. **X in GCX means all four RNA bases.

AMMONIA: a colourless, pungent gas (formula: NH_3), soluble in water, forming alkaline ammonium hydroxide, NH_4OH. Salts of ammonia have been known from the earliest times. The Romans mined *ammonium chloride* deposits near the temple of Jupiter Ammon in ancient Libya and referred to the product as **sal ammoniacus**. In the modern world, ammonia has many industrial uses. The Haber-Bosch process for ammonia synthesis, invented in 1909, involves reaction of nitrogen and hydrogen at high pressure and temperature in the presence of an inorganic catalyst, these extreme conditions being necessary to overcome the high energy of the triple bond joining the two atoms of the nitrogen molecule. The annual world production of ammonia by this method is over 100 million tonnes, mainly used as fertilizer in agriculture. However, most global nitrogen fixation is biological, carried out by *diazotrophic* bacteria, literally 'feeding on N_2'. An important genus is that of the *Rhizobia* (**rhiza** Gk. 'root'), soil bacteria which form *symbiotic* nodules on the roots of *leguminous* plants. Nitrogen fixation is carried out by the *nitrogenase* complex, a *metalloenzyme* containing iron and molybdenum, using ATP as the energy source. The reaction of nitrogen and hydrogen to form ammonia is *exergonic*, i.e. energy-yielding, but has a very high *activation energy*, i.e. energy has to be supplied in the form of heat or ATP hydrolysis to initiate the reaction.

ANALGESIA: an- 'without' + **algos** (Gk.) 'pain'. Depending on context, analgesia may refer to an inability to feel pain or to the relief of pain. ANALGESIC: a drug that relieves pain. HYPERALGESIA: extreme pain MYALGIA: **mus** 'muscle', hence muscle pain. NEURALGIA: **neuron** 'nerve', so nerve pain. NOSTALGIA: **nostos** 'homecoming' + **algos**: invented in 1688 by a physician to describe the clinical symptoms of homesickness, hardly a clinical symptom in current usage, confirming the line 'nostalgia ain't what it used to be.'

ANTHROPO-: anthropos (Gk.) 'human being'.
ANTHROPOCENE: **anthropos** + **kainos** 'recent', a recent coinage in which the current geological age is defined as beginning with the Industrial Revolution because of its impact on the global environment. But not everyone agrees; some think it started about 11,500 years ago, coinciding with the beginning of agriculture, which also marks the beginning of the Holocene. Others have argued for a date about 60,000 years ago, when

humans spread from Africa, but usage favours its beginning with the Industrial Revolution.[14]

ANTHROPOGENIC, i.e. man-made, as in *climate change*.[15]

The corresponding stem in Latin is: HOMO (L.) 'man, mankind' as in *Homo sapiens*, literally 'wise man'—not to be confused with the Greek prefix **(h)omos** meaning 'the same' as in homogeneous. The *homo* stem is the basis for the classification of primates (see below), as well as the derivation of *human, humane, humanitarian, humanity*.

HOMINIDS and HOMININS: the *Hominidae*, also known as the great apes, are a taxonomic *family* of primates including four *genera*: chimpanzees and bonobos (*Pan*), gorillas (*Gorilla*), humans (*Homo*), and orangutans (*Pongo*). This classification is relatively recent. Previously the *Hominidae*, i.e. hominids, was the name given to humans and our extinct relatives, with the other great apes being placed in a separate family, the *Pongidae*. Humans and our extinct relatives such as *Homo erectus* are now referred to as *hominins*, members of the sub-family *Homininae*. For the taxonomically inclined, our complete current description is kingdom *Animalia*, phylum *Chordata*, class *Mammalia*, order *Primates*, family *Hominidae*, sub-family *Homininae*, genus *Homo*, of which *Homo sapiens* is the only extant species.

ANTHROPOID: resembling a human being. The term is often applied to apes but may also refer to all higher primates.

ANTHROPOLOGY: the study of human beings including cultural, physical, and evolutionary aspects.

ANTIBIOTIC: **anti-** + **bios** (Gk.) 'life'. Originally defined in 1942 by Waksman, the discoverer of *streptomycin*, as any compound produced by a microorganism that, even in high dilution, antagonizes the growth of other organisms. Many of the antibiotics in use today are *semisynthetic*, having been modified to improve their effectiveness. The original antimicrobials, the *sulphonamides*, developed in the 1930s, are completely synthetic. It has been suggested that it was the success of the sulphonamides that changed the attitude of medical researchers towards *chemotherapy*. This helps to explain why Alexander Fleming, who discovered *penicillin* in 1928, appeared to lose interest in it, working as he was at St Mary's Hospital in London, where the received wisdom at that time was that vaccines held the key to dealing with infectious diseases. Antibiotics are classified as either *bactericidal* or *bacteriostatic*, depending on whether they kill or simply inhibit growth. An increasingly serious problem and a potentially catastrophic one is the development of resistance to antibiotics due to their misuse and overuse. Compounding this problem is a decline in the rate of discovery and development of new classes of antibiotics.

ANTIOXIDANTS: Haemoglobin transports oxygen to all tissues where it is essential for metabolism. Oxygen, however, is also a highly reactive molecule that can damage cell constituents because of the formation by metabolic side reactions of what are termed reactive oxygen species (ROS). Antioxidants serve to limit this damage by being oxidized themselves. Antioxidants include vitamins A, C, and E together with *glutathione* and the protein *thioredoxin*. In addition, enzymes such as *catalase, superoxide dismutase* (SOD), and various *peroxidases* function to limit *in vivo* concentrations of ROS. ROS species include hydrogen peroxide (H_2O_2), *hypochlorous acid* (HClO), and free radicals such as the *hydroxyl radical* $\cdot OH$ and the *superoxide anion* O_2^-. One might expect that catalase, which breaks down H_2O_2, would be an essential enzyme, but people with a genetic deficiency are not adversely affected. However, mice lacking one of the SOD isozymes die soon after birth.

Vitamin E consists of a group of related lipid-soluble antioxidants, the *tocopherols*, which are yellow oily liquids. Derivation: **tokos** (Gk.), offspring + **pherein**, to bear, because a symptom of vitamin E deficiency is *infertility*, along with *ataxia* and *muscle wasting*. The predominant form of vitamin E is α-tocopherol, which prevents damage to cellular lipids containing *polyunsaturated fatty acids*.

Glutathione (GSH) consists of a tripeptide, γ-GluCysGly, in which cysteine and glycine are joined by a conventional peptide bond while the amino group of cysteine is joined to the γ-carboxyl group of glutamic acid. See also **AMINO ACIDS**. It serves as an *antioxidant* by virtue of its free thiol (-SH) group. Several *glutathione peroxidases* (Gpxs) have been characterized that contain the unusual amino acid selenocysteine (Sec) at their *active sites*—see **SELENIUM** entry. Gpx1 is expressed in all cell types. Gpx2 and 3 are restricted to particular cell types. Gpx4 is specific for lipid peroxides and thus complements the action of vitamin E. Gpx4 gene knockouts lead to embryonic lethality, and this likely accounts for at least part of the pathology associated with selenium deficiency.

Each of these enzymes catalyse concerted reactions that lead to the oxidation of glutathione. This results in formation of a disulphide bond between two glutathione molecules to give GS-SG. It can be reduced back to GSH by glutathione reductase with NADPH the *reductant*. The ratio GSH/GSSG within cells is usually high, but its variation can be used as an indicator of *oxidative damage*. The importance of GSH as an antioxidant is indicated by the finding that *knockout mice* unable to synthesize GSH die soon after birth.

Another essential antioxidant is *thioredoxin*, which upon oxidation is reduced again by glutathione-dependent thioredoxin reductase. Thioredoxins are present in all known organisms. In humans, they are encoded by two genes, TXN and TXN2. Loss of function of either of these genes is lethal at the four-cell stage of the developing embryo. The likely reason for this is that thioredoxin is the source of reducing power in the reaction carried out by *ribonucleotide reductase* that converts ribonucleoside diphosphates to the corresponding *deoxyribonucleosides*. Thioredoxin is a small protein characterized by the presence of two *vicinal* cysteines in a CysXXCys *motif*. Transgenic mice that had been modified to overexpress thioredoxin were found to live 35 per cent longer than the control mice, in support of the theory that free radicals contribute to ageing.

APICOMPLEXA: a *phylum* containing several thousand *parasitic protists*, characterized by a complex of *organelles* at the *apical* end (**apex** L. 'tip or pointy end') visible only by electron microscopy. This structure is also referred to as an *apicoplast*. This displays *homology* with the chloroplasts of plants and algae and is believed to have derived from an alga by secondary *symbiosis*. Apicomplexa are therefore thought to have had photosynthetic ancestry, which was lost as they adapted to animal *parasitism*. Some apicomplexans have subsequently lost the apicoplast, such as the human parasite *Cryptosporidium*. However, *Plasmodium* species which are responsible for the major human disease *malaria* have retained the apicoplast, which in this case is essential for survival of the parasite. Apicomplexans were formerly part of the *Sporozoa* phylum but now constitute a separate phylum *Apicomplexa*, all members of which are *obligate parasites* of animals.

SPOROZOITE: a cell form exhibited by Apicomplexans, examples being *Plasmodium* and *Babesia*, which are the agents causing *malaria* and *babesiosis* respectively. The sporozoite is the form which infects new hosts. In the case of *Plasmodium*, it multiplies within mosquito salivary glands and enters the human bloodstream as the mosquito bites and later the liver, where further multiplication occurs. See also entry for **SPORE**.

ARCHAEA: from **arkhaios** (Gk.) 'ancient' and **BACTERIA**: from **bakteria** (Gk.) 'staff or cane', because the first bacteria seen under the microscope were rod-shaped. Living organisms were originally classified as belonging to one or other of two domains, either *prokaryotes* or *eukaryotes*, the former corresponding to bacteria lacking a *nucleus* (**karyon** Gk.). Carl Woese (1928–2012) pioneered the study of *bacterial phylogeny* (**phulon** Gk. 'race') by comparing their *ribosomal RNAs*, these being suitable for comparisons extending over long periods of evolutionary time because of their low mutation rate, their role in protein synthesis having remained constant. This led to the discovery, published in 1977, of *archaebacteria*, now called *Archaea*, prokaryotes which in many respects are as distantly related to true bacteria as bacteria are to eukaryotes. All living organisms are therefore now regarded as belonging to one of three domains of life, *Eubacteria, Archaea*, and *Eukaryota*. Many Archaea are found in environments with extremes of pH and temperature; they are *extremophiles*. Traditionally the study of microbes has relied on the ability to grow pure cultures in the laboratory, but it has become apparent that only a small proportion of prokaryotes has ever been successfully cultured. A partial solution consists of genomic sequencing of single cells; another approach involves *metagenomics* where sequencing is applied to a mixture of cells in which particular species are dominant. These approaches are able to yield information about gene content and have recently been applied to uncultured *microbial dark matter*. Results of this work include evidence for new microbial phyla and examples of *lateral gene transfer* between Eubacteria and Archaea. While it has long been assumed that eukaryotes developed from simpler prokaryotes, as both their names suggest, there has until now been a complete lack of evidence for organisms that might represent evolutionary intermediates. Great interest therefore attaches to recent findings that Archaea obtained from marine sediments in the vicinity of *hydrothermal vents* contain genes that encode a variety of proteins previously found only in eukaryotes, so-called *eukaryotic signature proteins*.[16]

ASCORBIC ACID: **a-** 'not' + **scorbutus** (L.) 'scurvy'. Also known as vitamin C because humans, along with a few other vertebrates, are unable to synthesize it. Vitamin C deficiency is the cause of scurvy. The reason for this is that vitamin C is a *cofactor* for the *hydroxylase* enzyme that converts proline residues to *hydroxyproline*, one of the post-translational steps in the synthesis of the important structural protein *collagen*. Ascorbic acid is also an *antioxidant*. It reacts with *free radicals* or REACTIVE OXYGEN SPECIES (ROS), which are known to damage nucleic acids, proteins, and lipids. It is also reported to be required by TET enzymes that convert methylated C residues in DNA to hydroxylated derivatives. The discovery of vitamin C was largely due to the work of Albert Szent-Gyorgyi in the 1930s. The story is told that he submitted a paper for publication reporting its isolation and that its structure appeared to be similar to that of a sugar but that, in spite of extensive investigation, its exact structure was still unknown. He therefore suggested the name ignose. The journal editor thought this name was too frivolous and asked him to provide an alternative. He replied, suggesting godnose. Szent-Gyorgy was a prominent figure in the resistance to the Nazis in his native Hungary and had the distinction of being the subject of a personal order from Hitler for his elimination, an outcome that he evaded.

AUTOIMMUNE DISEASES. More than fifty different diseases affecting 5 per cent of the population are now known to be due to the immune system reacting against particular self-antigens (Davis 2018). One example is *coeliac disease*, also called *gluten-sensitive enteropathy*, where sufferers need to avoid *gluten*, a protein present in most grains. Susceptible individuals are a subset of those that express *human leukocyte antigen* (HLA) genes DQ8 or DQ2. Another autoimmune disorder is *narcolepsy*, shown to be caused by the elimination of a population of cells within the *hypothalamus*—see the entry for NARCOLEPSY under **SLEEP**. The cause of autoimmunity is not clear. One possibility is that among the range of *antibodies* elicited by a *pathogen*, especially under conditions of a strong immune response, there will coincidentally be those that have activity against particular host proteins/antigens (cross-reactivity). Another possibility is that the immune mechanisms that distinguish between *self* and *non-self antigens* may not always be 100 per cent efficient. Some evidence for this comes from work on *cancer immunotherapy*, where it has proved possible to manipulate the strength of the immune response.[17]

BACILLUS: bacillus (L.) 'small rod', diminutive of **baculum** 'rod'. The term *bacillus* can refer to any rod-shaped bacterium. *Bacillus*, capitalized and italicized, refers to a particular genus of *gram-positive*, spore-forming bacteria which contains over 200 species. The type species is *B. subtilis*, which has served as a model organism for investigation of many facets of bacterial molecular biology.

BILE: khole (Gk.) 'bile'. A dark-green liquid produced by the liver and stored in the gall bladder, its function being to solubilize lipid components of the diet, thereby promoting their digestion and absorption. This is accomplished by the *bile salts*, mainly *glycocholate* and *taurocholate*, *amphiphilic* molecules with strong detergent action that consist of the ring system of cholesterol, to which are attached polar side chains. These amount to 10 per cent of bile. See also entry for BILE SALTS under **CHOLESTEROL**. *Cholesterol* is also present in bile at low concentration. Other components of bile include *choline* as well as *biliverdin* and *bilirubin*, breakdown products of haem.

The Greek physician Galen (CE 129–ca. 200) was a proponent of the idea that went back at least to Hippocrates in the fifth century BCE that human personality traits depended on the balance between four bodily humours (blood, phlegm, yellow bile, and black bile) and that an excess of one or the other was the basis of disease. In medical circles, this continued to be gospel until well into the sixteenth century. An echo of this teaching can be found in the following English words: *sanguine*, meaning, according to the *OED*, someone who is characterized by a ruddy countenance and a courageous, hopeful, and amorous disposition, from **sanguis, -inis** 'blood'; *phlegmatic*, meaning someone who is unemotional or even apathetic, from **phlegma**, which in Greek meant a morbid humour but which the Romans translated as 'a clammy humour of the body'; *choleric*, meaning someone who is bad-tempered or irritable, from **khole**, i.e. bile; the word *cholera* has a similar derivation but in English now refers to the life-threatening disease. At the school I attended, a cranky teacher was said to be afflicted, not with khole, but with 'sol', which translated as 'shit on the liver'—same idea. *Melancholy* or **melancholia**, meaning a gloomy state of mind or depression, derives from **melanos** 'black' + **khole**. The attitude of the Greeks to melancholy was not entirely negative. Aristotle refers to 'the melancholy of exceptional men'. 'Why is it', he asked, 'that all men who have excelled in philosophy, in politics, in poetry or in the arts, have been subject to melancholy?' Autumn was its season, dusk its time of day, earth its element, and Saturn its planet. Melancholy would be diagnosed today as depression. The word *bilious* can mean cross or annoyed as well as sick or nauseated. A related word is *atrabilious*: from **atra bilis**, 'black bile', being the Latin translation of the Greek **melancholia**.

-BLAST and BLASTO- blastos (Gk.) 'sprout'. Words containing the prefix or suffix 'blast' can refer to *proliferating precursor cell lineages* as well as to cells that are *terminally differentiated* and activated. *Fibroblasts*, the most common cells in *connective tissue* and which synthesize components of the *extracellular matrix* such as *collagen*, are examples of terminally differentiated cells, which may or may not be capable of being activated. Active fibroblasts are distinguished by their abundant **ENDOPLASMIC RETICULUM**. *Osteoblasts*, bone-forming cells, are said to be terminally differentiated and, as such, synthesize a variety of proteins that constitute the organic matrix of bone. However, individual osteoblasts are only able to lay down the calcium- and phosphate-based minerals of bone when part of an organized group of cells called an *osteon*.

'Blast' words also designate stages and structures of early *embryo* development (**embruon** Gk.: from **em-** 'into' + **bruein** 'to swell, grow'). The *epiblast* is the mammalian embryonic tissue that contains the *pluripotent stem cells* that generate the whole embryo. Fertilization is followed by multiple cell divisions and results in a ball of cells that then transforms into a hollow sphere, the *blastula*. These cells, distributed on the surface of the sphere, are referred to as *blastomeres*, while the fluid-filled cavity is the *blastocoele*, **blastos** + **koiloma**: 'cavity'. In mammalian embryos, the blastula is known as the *blastocyst*, similar to the blastula in other animals except that it contains what is referred to as the *inner cell mass* (ICM), located within and to one side of the cavity. It is the ICM, also called the *embryoblast*, that goes on to form the *fetus*. The cells on the surface of the blastocyst are known as the *trophoblast*. The trophoblast embeds the blastocyst in the *endometrium* of the uterine wall and goes on to form part of the *placenta*. *Tropho-* means food or nutrient, the supply of which to the developing embryo is a role the trophoblast fulfils from the blastula stage on. For figures and additional information concerning *blast-* etc., see https://en.wikipedia.org/wiki/Blastocoel.

BLOOD (OE) consists of a liquid component containing the soluble *blood proteins* and a cellular component that includes the haemoglobin-containing red blood cells or *erythrocytes*, **eruthros** (Gk.) 'red'+ **kutos** 'vessel', together with a variety of white blood cells or *leukocytes*, **leukos** (Gk.) 'white', and *platelets*.

Blood *plasma* is the liquid that is obtained by removing the cellular component and *fat globules* by centrifugation. The liquid that separates out when blood clots is referred to as *serum*. The main difference is that serum does not contain *fibrinogen*, which is converted to the *fibrin* clot. The Greek word for blood was **aima**, transliterated as *haem* in English, *heme* in American; cf. **sanguis** (L.), also meaning blood but with few descendants in English *sanguine*, 'cheerful'; *sanguinary*, 'bloodthirsty', *consanguineous*, 'related by birth', i.e. a blood relative.

HAEM is the name of the *tetrapyrrole* compound which, with a *ferrous* ion at its centre and when bound to its globin *apoprotein* to form *haemoglobin*, carries oxygen to the tissues. But *haem-* also serves as a combining form in a large number of other words that relate to blood, some of which are listed here.

ANAEMIA: usually refers to a low red blood cell count, which can result from a variety of causes.

HAEMODIALYSIS: the life-saving procedure for removing toxins and metabolites from the blood when the kidneys are not working.

HAEMAGGLUTININS: **agglutinare** (L.) 'to fasten to' from **gluten** (L.) 'glue': proteins which cause clumping of red blood cells. Examples are found on the surface of the flu virus. Strains of flu are characterized as H1N1 etc., the H standing for haemagglutinin and N for the enzyme *neuraminidase*.

HAEMOPROTEINS: proteins containing one or other of a variety of haem groups, such as haemoglobin, but also including the *cytochromes* of the electron transport chain and the *P450 cytochromes*.

HAEMORRHAGE: blood loss usually indicating uncontrolled bleeding, the suffix here from the Greek meaning 'to burst'.

HAEMOSTASIS: the prevention of bleeding: **stasis** (Gk.), 'stoppage'.

HAEMATOLOGY: the study of blood and associated diseases.

ISCHAEMIA: from **iskhaimos** 'stopping blood' from **iskhein** 'to keep back', usually indicating an inadequate supply of blood to the muscles of the heart.

LEUKAEMIA: a group of cancers affecting white blood cell production.

SEPTICAEMIA: from **septicus** (L.) and **septein** (Gk.) 'to putrify': blood poisoning, usually due to a bacterial infection.

THALASSEMIA refers to anaemias caused by mutations in the α- and β-subunits of haemoglobin which prevent normal levels of haemoglobin synthesis. The name comes from **thalassa** (Gk.), meaning sea, because these genetic diseases were first characterized in countries surrounding the Mediterranean.

BLOOD PROTEINS: *albumin*: **albus** (L.) 'white', **albumen** (L.) 'the white of an egg', the major protein of which is now referred to as ovalbumin. *Serum albumin* is the major protein in blood plasma (55 per cent) and plays an important role binding and transporting in the blood a range of low-molecular-weight hydrophobic molecules of limited solubility in water. In the early days of protein chemistry, many proteins were classified as either albumins (soluble in pure water) or *globulins* (soluble in dilute salt solutions but insoluble in pure water). Blood contains a complex mixture of proteins, many in minute amounts. The traditional separation is that provided by *electrophoresis*, where in addition to albumin, in order of decreasing mobility, are α-, β-, and γ-*globulins*. The γ-globulins or *immunoglobulins* have similar overall architecture but are extremely heterogeneous. Another twenty or more proteins belong to the *complement system*, part of the *innate immune system* that provides protection against microbial invaders and, as the name suggests, complements the antibody system. Also among the soluble blood proteins are the numerous clotting factors (see below), including *fibrinogen*, which accounts for 7 per cent of plasma proteins. While haemoglobin is present at high concentration within red blood cells, only in *haemolytic anaemia* is free haemoglobin released into the blood.

Free haemoglobin forms a high affinity complex with the blood protein *haptoglobin*, named from **haptein** (Gk.) meaning 'to fasten', which is then degraded in the spleen. Another protein is *haemopexin*, **haima** + **pexis** 'putting together', a protein that binds haem with very high affinity and appears to serve two functions: to conserve iron and to prevent oxidative damage that could result from free haem. It releases iron upon binding to a receptor on the surface of liver cells.

BLOOD CELLS are conveniently divided into red ones that transport oxygen to the tissues and white ones that are part of the immune system.

HAEMATOPOIESIS: (**haemato-** (Gk.) 'relating to blood' + **poiesis** 'making, producing', i.e. blood production, involves the differentiation of an initial *haematopoietic stem cell* (HSC) into red and white blood cells. This occurs in the bone marrow, where both red and white blood cells and other cells of the immune system derive from the common *myeloid* precursor (**muelos**: 'marrow'), the HSC. Thus *erythroblasts*, nucleated cells, are precursors to *erythrocytes*, red blood cells, which are *anucleate*, while *myeloblasts* give rise to white blood cells: *basophils, neutrophils, eosinophils,* and *monocytes*. The HSC also gives rise to a *common lymphoid precursor*, the precursor of *B-cells*, *T-cells*, and other lymphoid cells via a precursor *lymphoblast*. It is estimated that there are between 10,000 and 20,000 haematopoietic stem cells in human bone marrow, which produce 100 billion (10^{11}) new blood cells every day. See also entries for **STEM CELLS** and **IMMUNITY**.

ERYTHROPOIESIS, also HAEMOPOIESIS: the process whereby red blood cells are produced. This process is sensitive to oxygen levels which, when low, cause the kidney to secrete the hormone *erythropoietin*, which stimulates the proliferation and differentiation of red blood cell precursors. Athletes who compete in endurance events often train at high altitudes in order to stimulate red blood cell production and hence their oxygen-carrying capacity. The advent of recombinant erythropoietin initiated a short-lived attempt by less-scrupulous athletes to enhance their performance using the recombinant product.[18]

BILIRUBIN: **bilis** + **ruber** 'red' and BILIVERDIN: **bilis** + **verde** 'green'. The human red blood cell has a life span of about 120 days. Old cells are degraded in the *spleen*. While the iron of haemoglobin is recycled, the globin is broken down to amino acids and the haem group is converted to a *linear tetrapyrrole, biliverdin*, by *haem oxygenase*, a member of the *cytochrome P450* superfamily. Biliverdin is then reduced to bilirubin by *biliverdin reductase*. Bilirubin is not very soluble in water. Bound to serum albumin, it is transported to the liver, where it is converted to the more soluble *glucuronide* derivative and secreted into the bile.

LEUKOCYTES—WHITE BLOOD CELLS: **leukos** (Gk.) 'white'

Like red blood cells, white blood cells are made in the *bone marrow*. The myeloid leukocytes are part of the *innate immune system*, which acts immediately to eradicate infections, as distinct from the antibody-based *adaptive immune system*, which can take a week or more to respond to an antigen not previously encountered. The myeloid cells include *polymorphonuclear* leucocytes, so called because their nuclei have irregular lobular shapes. Polymorphonuclear white blood cells typically contain secretory granules in their cytoplasm and are classified as *granulocytes*, from **granulum**, the diminutive of **granum** 'granule or seed'. Three types are distinguished on the basis of their staining behaviour. *Basophils* are the least common white blood cells, 0.1–0.3 per cent. These stain blue with basic stains. *Eosinophils* stain red with acid stains. *Neutrophils* stain neutral pink with eosin. Another type of myeloid cell is the *monocyte*, 2–10% of the total, which does not have multiple lobes in its (non-polymorphic) nucleus. Monocytes differentiate into either *macrophages* (see also under **PHAGE**) or *dendritic cells*, from **dendron** (Gk.) 'tree', which function as *antigen-presenting cells* and can leave the blood and enter the tissues.

Neutrophils are the most abundant white blood cell, accounting for 60—65 per cent of the total at a concentration of 5×10^9 cells per litre of blood. Their granules contain a cocktail of bactericidal enzymes including proteases, lysozyme, *myeloperoxidase*, and an iron-binding protein. Following an infection, neutrophils leave the blood and can reach the site of infection within thirty minutes. These neutrophils do not return to the blood but undergo apoptosis, releasing their cargo of toxic chemicals, and form pus. Neutrophils are also attracted to so-called *sterile inflammations* caused by trauma, toxins, or atherosclerosis, and in this case, having contributed to tissue repair, they may return to the bone marrow and then undergo apoptosis.[19]

MEGAKARYOCYTE: **mega** 'large' + **karyo-** 'nucleus' + **cyte** 'cell'. These are also derived from haematopoietic stem cell precursor cells in bone marrow. As their name implies, they have very large nuclei because their DNA goes through many rounds of DNA replication without initiating cell division (*cytokinesis*). Thus, megakaryocytes may contain up to thirty-two times the normal cellular DNA content. Megakaryocytes are ten to fifteen times the size of red blood cells and are not normally seen in peripheral blood because they are too large to pass through *capillaries*. In the bone marrow, each megakaryocyte gives rise to 2,000–5,000 *platelets* or *thrombocytes*. Platelets play an essential role in blood clotting—see below.

CYTOPENIA: less than the normal number of blood cells, from **cyto-** + **penia** (Gk.): 'poverty or need'. Depending on the type of cell, we can have *leukopenia, lymphopenia, thrombocytopenia, granulocytopenia,* or *pancytopenia,* i.e. the whole lot, but a reduced red blood cell count is *anaemia*.

BLOOD VESSELS. The Latin for vessel is **vas, vasa** with the diminutive **vasculum** 'small vessel', from which we get *vasculature,* a general term as in 'tumour vasculature' and the adjective *vascular* as in 'tissues that are metabolically active are highly *vascular*'. While both Greek and Latin retained the idea of *vessel* as a receptacle, English has vessels meaning the networks that carry blood and *lymph* around the body. Some liver cells, i.e. *hepatocytes,* differentiate into *bile duct cells,* known as *cholangiocytes,* **khole** (Gk.) 'bile'. Linnaeus, who was writing in Latin, retained the original meaning when he defined *angiosperms* as those plants having enclosed seeds, as distinct from the *gymnosperms,* **gymnos** 'naked' + **sperma** 'seed'. Greeks also had **phleps, phleb-** meaning 'vein' and **phlebotomia** 'blood-letting', *phlebotomy* in English, a favourite panacea practised by physicians in the Middle Ages that continued till the nineteenth century. *Phlebitis* is inflammation of the walls of a vein.

ANGIOGENESIS or ANGIOPOIESIS: **angio-** relating to blood vessels, from **angion** (Gk.) 'vessel or receptacle'. Blood vessel formation depends on a group of related growth factor proteins that promote formation of blood vessels. It was suggested that inhibitors of *angiogenesis* might be useful treatments for solid tumours which require additional blood vessels for their growth, but this approach has not been successful. It appears now that better blood flow gives better access for both chemotherapy and immunotherapy. For ANGIOGENESIS, see also under AUTOCRINE heading in **SECRETION MECHANISMS**.

ANGIOTENSIN is an oligopeptide hormone which causes constriction of blood vessels, i.e. *vasoconstriction*, from **constringere** (L.) 'to bind tightly together', and so increases blood pressure. It is derived from a large precursor protein, *angiotensinogen*, by two steps of proteolysis, initially by *renin*, which the kidney secretes in response to decreased blood pressure. The second proteolysis is by *angiotensin converting enzyme* (ACE). ACE also raises blood pressure by degrading a potent *vasodilator*, *bradykinin*. ACE is therefore a target for drugs that treat *hypertension*. A variant of ACE, ACE 2, has acquired sudden notoriety as the receptor for the coronavirus responsible for the COVID-19 pandemic.

ARTERIOSCLEROSIS: **arteria** 'artery' + **sklerosis** 'thickening'. Healthy arteries are flexible and elastic. Over time, high blood pressure can reduce blood vessel elasticity, thereby restricting blood flow to organs.

ATHEROSCLEROSIS: based on German **atherosklerose**. A type of arteriosclerosis characterized by the accumulation of lipids and fibrous plaques in swellings termed *atheromata* located within the *intimal layer* immediately adjacent to arteries. Disruption of the atheromata results in thromboses that, depending on location of the artery, may cause heart attack, stroke, or other organ failures.[20]

ANEURYSM: **aneurunein** (Gk.) 'to dilate'. A balloon-like bulge in the wall of a blood vessel, usually an artery.

BLOOD-BRAIN BARRIER (BBB): *Endothelial cells* that line the blood vessels of the brain prevent large polar molecules from entering or leaving, isolating brain metabolism from other tissues by excluding some hormones and metabolites. Special transporters provide access for glucose, amino acids, fatty acids, and some other small molecules. The BBB also excludes most bacteria from the brain, although some, such as those causing meningitis, force entry by degrading the endothelial barrier. Most drugs are also excluded by the BBB. The possibility of using *liposomes* loaded with *nanoparticles* is being investigated as a method for drug delivery to the brain.

CAPILLARY: **capillus** (L.) 'hair'; smallest blood vessel. Capillaries are lined with a single-cell layer of endothelium. Small gaps between adjacent endothelial cells allow molecules from the bloodstream to pass into and out of surrounding tissues.

THROMBOSIS: local, pathological formation of a blood clot (from **thrombos** Gk. 'lump, blood clot'). While activated platelets are central to haemostasis, they are also implicated in the occurrence of *thromboses* in arteries that supply heart, brain, and other vital organs. Deep vein thrombosis is a potential risk associated with long air flights. EMBOLISM, caused by an embolus, **embolus** (Gk.) 'stopper', is an obstruction in an artery caused by a mobile object, usually a blood clot, air bubble, or dislodged arterial plaque. The obstruction can interrupt downstream blood supply to produce an INFARCT, a localized area of dead tissue, from **infarcire** (L.) 'to stuff with something'.

BLOOD CLOTTING occurs in two stages. First, *platelets*, the *anucleate* cells derived from *megakaryocytes*, are activated to plug the site of blood vessel injury. Then a fibrin clot forms from *thrombin* action on *fibrinogen*. Thrombin is a serine protease belonging to the chymotrypsin family and plays a central role in *haemostasis*. Thrombin is synthesized as the inactive *zymogen* precursor *prothrombin*. Several glutamic acid residues near the N-terminus of prothrombin are converted to γ-*carboxyglutamic* acid in *vitamin K*-dependent reactions. Those residues strongly bind calcium ions. This in turn causes prothrombin to bind to phospholipid surfaces, where it is converted to thrombin by *prothrombinase*. For more on vitamin K and the occurrence of γ-carboxyglutamyl residues, see the entry for OSTEOCALCIN. Prothrombinase is a complex consisting of factors, including factor Xa, produced by a cascade of zymogen activations that are set in motion by tissue factors at the site of injury. Factor Xa can convert prothrombin to thrombin independently, but the assembled prothrombinase complex accelerates the reaction by several orders of magnitude. Mutation of either of two of the factors on this pathway, factor VIII or factor IX, result in the bleeding disorder *haemophilia*. A final step in clot formation is when thrombin activates factor XIII, a *transglutaminase*. Transglutaminase strengthens the clot by cross-linking fibrin molecules via *isopeptide* bonds between the amide group of glutamines and side-chain amino groups of lysines, with the elimination of glutamine nitrogens.

$$-CH_2-CO-\underline{N}H_2 + NH_2-CH_2- \rightarrow -CH_2-CO-NH-CH_2- + \underline{N}H_3$$
Glutamine   Lysine

CADAVER: (L.) the word in Latin can refer to 'cadaver, corpse, or carcase'. English discriminates between these alternatives as follows: cadaver would normally refer to a dead human body possibly subject to dissection in medical school; corpse also refers to a dead human body, while carcase means a dead animal but is sometimes employed humorously to refer to a live body (someone else's). Related words are *cadaverous*: having a pale corpse-like appearance; *cadavorous*, from **cadere** (L.) 'to fall' + **vorare** 'to devour', describes the behaviour of predators that feed on animal carcases. Then there is CADAVERINE or 1,5-diaminopentane formed by the decarboxylation of the amino acid lysine, a reaction which occurs during the putrefaction of biological material. Cadaverine is one of a small group of compounds known as polyamines that bind strongly to DNA and RNA.[23]

Corpus (L.), plural **corpora**, also means body, either alive or dead depending on context, hence *corpse, corporal punishment*, and the diminutive *corpuscle*, which can refer to red or white blood cells but which, like corporal punishment, has gone out of fashion; also *corpulent*, an adjective used to describe a body, a fat one. The Greek **nekros** also means 'corpse', from which various English words are derived, such as *necrosis*, meaning death of cells or tissue due to injury or disease and *tumour necrosis factor* (TNF), actually a family of cytokines that are mainly concerned with modulation of the immune response. Another term is *necroptosis*, which is a programmed version of necrosis. It can be thought of as a mechanism of *cell suicide* that operates to abort infection. It is restricted to vertebrates and possibly developed in order to cope with viral infections when *apoptosis* is blocked by the virus.

The derivation of cadaver is from the verb **cadere, cecidi, casum** (L.) meaning *to fall, sink*, or *die*. A large number of English words derive from verbs related to **cadere**: *excise* from **excidere**, *concise* from **concidere**, *accident* from **accidere**, *incident* from **incidere**, while **decidere**, which means to fall down, has given rise to *deciduous*, *decide* as well as *decay* via OF **decair**, the *Occident* from **occidere**, meaning the West, i.e. where the sun falls below the horizon; cf. the *Orient*, i.e. the East, from the verb **oriri**, (L.) 'to rise'; the **or-** stem + the **-igo** suffix also gave rise to **origo** (L.) meaning 'origin'. Another related verb is **caedo, caedere, cecidi, caesum**, meaning 'to cut or kill', the root for the English suffix *-cide* and thus *homicide*, *herbicide*, *insecticide*, etc. The same past tense, **cecidi**, meaning both 'died' and 'killed' may say something about the way Rome operated. In Shakespeare's *Julius Caesar*, he has Mark Antony say:

> For Brutus, as you know, was Caesar's angel:
> Judge, O you gods, how dearly Caesar loved him!
> This was **the most unkindest cut of all**;

Evidently, Shakespeare here is indulging in a *bilingual pun* involving Caesar's name and the Latin past participle for cut, **caesum**. (One commentator has suggested, with tongue in cheek, that the term *bilingual* should be replaced by *translingual*, on the grounds that the word *bilingual* implies intimate physical contact.) Also, the assertion that Julius Caesar was born by caesarean section seems unlikely.

CANCER: cancer (L.) The word in Latin originally meant 'crab' and later came to mean 'cancer'. Derivation of the term for the medical condition came about because the swollen veins surrounding tumours resembled the limbs of a crab. Galen, writing in the second century CE, quoted by Mattern (2013): 'We have often seen in the breasts a tumour exactly similar to that animal, the crab. For just as in the crab the feet are on either side of the body, so also in this disease the veins, extending from the unnatural tumour, make a shape similar to a crab.' Mattern continues, 'Galen believed the condition was caused by an accumulation of black bile not adequately cleared through menstruation, and that he could cure it in its early stages with purgative drugs'.[21] Lewis and Short define **cancer** in its medical context as 'a crawling, eating, suppurating ulcer, malignant tumour'. A more recent definition is 'Cancer is a disease of uncontrolled cell division that is fuelled by genetic instability'. The corresponding word in Greek is **karkinos**, hence *carcinoma*: a cancer that begins in tissue that lines the inner or outer surfaces of the body. Also, *carcinogen*, an agent that causes cancer which includes many chemicals as well as the ultraviolet component of sunlight which gives rise to skin cancers such as *melanoma*.

A major portion of current biological research is concerned with cancer, and much progress has been made since Galen's day in understanding the causes of the disease as well as its treatment. Recent advances justify optimism that its impact can be further alleviated. Most cancer deaths occur as a result of *metastasis*, which occurs when a primary lesion spreads to other sites, so it follows that methods that allow early detection are needed. It has recently been shown that DNA sequences characteristic of some tumours as well as tumour cells themselves can be detected in blood samples. Such 'liquid biopsies' enable earlier detection as well as a means of assessing the progress of treatment.[22] A major problem in dealing with any particular cancer is that as it grows, so it evolves, resulting in resistance to both *chemotherapy* and *immunological surveillance*. See also the entry for **ORGANOIDS** for a recent method for optimizing chemotherapy.

ONCOGENES. The prefix ONCO- from the Greek **onkos** meaning 'mass or swelling' relates to tumours, with ONCOLOGY being that branch of medicine that deals with cancer. An ONCOGENE is a normal cellular gene which when expressed inappropriately, that is to say overexpressed or expressed at the wrong time, may result in formation of a tumour. An ONCOPROTEIN is the protein product of an oncogene. Oncogenes typically are part of *signal transduction* pathways such that their *inappropriate expression* results in *uncontrolled cell growth*. Situations that can cause a gene to become oncogenic include *chromosome translocations, incorporation into a viral genome*, or *mutations occurring in sequences controlling its expression*. In each case, the result is loss of normal control of expression of the gene.

The first oncogene to be characterized was the *src* gene, a component of the Rous sarcoma virus, named for Peyton Rous (1879–1970), who in 1909 isolated it from a chicken sarcoma and showed that it was *transmissible*, i.e. able to induce a tumour in another chicken. It was not till 1976 that the *src* gene was isolated and the crucial discovery was made that it was a normal chicken gene which, during an infection at some time in the past, had become incorporated into the virus. Other oncogenes were soon characterized, and it became clear that they were endogenous genes that give rise to tumours as a result of a genetic mutation of one sort or another. A cellular gene with oncogenic potential is also called a *proto-oncogene*.

As well as oncogenes, there are genes known as *tumour suppressor* genes, which are often found inactivated by mutation in cancers, most commonly the *TP53* gene, which has been described as the guardian of the genome. *TP53* is the most frequently mutated gene in cancers. The protein encoded by *TP53* prevents tumour development by a number of mechanisms, including activation of DNA repair enzymes when DNA has sustained damage. It can also inhibit the cell cycle in order to buy time for DNA repair to occur and, if such damage is irreparable, trigger *apoptosis*, i.e. cell death. Its effects also extend to regulation of ammonia excretion via the urea cycle in cancer cells with consequent inhibition of *polyamine* synthesis and cell proliferation.[23] Numerous other tumour suppressor genes have also been identified.

CANCER GENOMES. The advent of *rapid DNA sequencing* has enabled analysis of large numbers of cancer whole genomes as well as matching genomes from healthy tissue. Recent papers by an international consortium report the results of a *meta-analysis* of 2,658 whole cancer genomes from thirty-eight different tumour types. These underline the complexity of cancer genomes that, having escaped from the normal controls of *cell division*, proceed to evolve in unpredictable ways.[24]

Tumour genomes may contain anywhere between 1,000 and 20,000 *somatic mutations*. These mutations are categorized as either *driver mutations*, i.e. mutations that contribute to the initiation and/or progression of the tumour, the remainder being *passenger* mutations which simply go along for the ride, many of which may predate the tumour. On average, each tumour contains four to five driver mutations, most of which occur in coding regions, the major exception being mutations in the promoter of the *telomerase* gene which occur in up to 71 per cent of *melanomas* and more than half of *bladder cancers* and *glioblastomas*. These mutations increase the activity of *telomerase*, allowing tumour cells to divide uncontrollably. Normal cells have finite cell division potential partly due to the accompanying gradual reduction in telomere length. Overexpression of the telomerase gene is one way that cancers achieve *replicative immortality*.

Estimates of the total number of potential drivers are in excess of 400. Examples of driver mutations that occur in some types of cancer are mutations that inactivate enzymes involved in the repair of DNA damage. Of driver mutations that involve *TP53*, 77 per cent have both alleles mutated with a point mutation in one allele and deletion of the second. This points to the frequent occurrence in tumour DNA of major structural variations. These include *inversions*, *translocations*, and *aneuploidy* resulting from widespread loss or gain of entire chromosomes. Many tumours also exhibit the results of what are described as genomic catastrophes. In the work referred to above, 17.8 per cent of tumours exhibited *chromoplexy*, **plessein** (Gk.) 'strike', which results from the occurrence of several double strand breaks which are then rejoined incorrectly. Another major genomic defect occurred in 22.3 per cent of tumours as a result of *chromothrypsis*, **thrypsis** (Gk.) 'a breaking into many pieces', in which tens to hundreds of DNA fragments are rearranged. This commonly occurs in melanomas.

CHRONIC MYELOID LEUKAEMIA (CML), also known as *chronic myelogenous leukaemia*, is an example of a cancer that can now be successfully treated in most patients. CML is characterized by the accumulation of immature white blood cells, known as *blast cells*. The final stage of CML is known as the blast crisis stage, where up to 30 per cent of the patient's white cells consist of blast cells. The cause of CML is a reciprocal *chromosome translocation* in which a portion of chromosome 22 is joined to chromosome 9 to generate what is known as the Philadelphia chromosome. The result is formation of a fusion gene called *bcr/abl*, which gives rise to a fusion oncoprotein (*bcr* stands for *breakpoint cluster region*). The *abl* gene encodes a *tyrosine kinase* enzyme, i.e. an enzyme that is able to phosphorylate the -OH group of a tyrosine side chain in an acceptor protein. Under normal conditions, this tyrosine kinase is part of a tightly controlled signalling cascade that induces the generation of white blood cells. Formation of the *bcr/abl* fusion gene causes loss of control of *abl* gene expression with resultant chronic overproduction of blast cells (Wapner 2013).[25]

A treatment for CML came with the synthesis of the tyrosine kinase inhibitor Gleevec, which underwent initial human testing in 1998. The worry at the time was that there would be side effects due to inhibition of other tyrosine kinases, but this was not the case. Wapner relates how at various stages in the research that led to the development of Gleevec, the bean counters argued against the project on the grounds that the number of CML patients was not sufficient to justify the expense, and in any case, such cancer patients tended to not survive for long. Gleevec disproved both arguments. More than 80 per cent of patients are surviving longer than ten years, and the predictable problem of development of resistance to the inhibitor has been managed by development of alternatives to Gleevec. The other major consequence of this work stems from the fact that tyrosine kinases are common enzymes and implicated in many different diseases. Gleevec provided proof of principle that tyrosine kinases could be effectively and specifically inhibited, and this has unleashed a major research effort in this area.

CANCER IMMUNOTHERAPY. Until recently, the only treatments for cancer have been surgery, radiation, and chemotherapy. This has changed with the rediscovery that the immune system can be stimulated in a manner that in some circumstances results in the elimination of tumours. The initial approach has been to use specific antibodies to remove or inactivate proteins that are bound to T-cells or dendritic cells and that act as brakes on the immune system. This procedure is referred to as *immune checkpoint therapy*. Most of the early trials on cancer patients targeted two such so-called *brake receptors*, CTLA-4 and PD-1. These work in different ways but show that immune cells *can* provide permanent cancer cures. The problem so far is that this approach only works for, in round numbers, one in five patients. The search is on to find *biomarkers*, i.e. characteristics of the patient or the tumour that would predict treatment outcomes. Improved outcomes will also depend on further modifications of T-cells that enhance their ability to target and kill cancer cells. These are early days, but there are reasons for optimism that the immune system can be further manipulated to improve survival of cancer patients.

CARBOHYDRATES, literally 'hydrated carbon', encompass a wide range of biological molecules containing carbon together with hydrogen and oxygen in a ratio of 2:1, e.g. glucose $C_6H_{12}O_6$, and dates from the early days of chemistry when *empirical formulae*, i.e. simple ratios of elements, were first determined. The word *sugar* is much older—humans have always had a sweet tooth—and derives from an ancient Indo-European root related to **sarkara** in Sanskrit, **sakkharon** in ancient Greek, and **saccharum** in

Latin. As a result, we refer to glucose as a *monosaccharide* and a polymer containing sugar molecules as a *polysaccharide*. Cane sugar only became common in Europe around the eighteenth century, and before then, sugar was a term applied to sweeteners in general, mainly plant extracts and honey. The Greeks became aware of cane sugar when Alexander journeyed to India but did not exploit it.

MONOSACCHARIDES. Sugars are designated by the suffix *-ose* and classified according to the number of carbons (*triose, tetrose, pentose, hexose,* and *heptose*), and depending on whether they contain an *aldehyde* group, the *aldoses*, or a *keto* group, the *ketoses*. Glyceraldehyde is a triose, with a single *asymmetric* carbon, or a single *chiral centre*, i.e. a carbon with four different *substituents*. This means that there are two possible arrangements of the substituents in three dimensions and therefore two possible *isomers* of glyceraldehyde, which are not *superimposable*. These are known as *stereoisomers*, from the Greek **stereos**, meaning 'solid', and are denoted by the letters D and L. D-glyceraldehyde and L-glyceraldehyde are mirror images of each other, also known as *enantiomers* from the Greek **enantios,** meaning 'opposite'. See also under entry for **CHIRALITY**. Most naturally occurring sugars are D-sugars. The convention is that a sugar with more than one asymmetric carbon is a D-sugar if the configuration at the asymmetric carbon furthest from the aldehyde group is the same as the configuration of D-glyceraldehyde. Aldoses, by virtue of the presence of the aldehyde group, are referred to as *reducing sugars*.

D-glucose (**glukus** in Greek means 'sweet') is the key sugar molecule in animals. The control of blood glucose concentration by *insulin* is critical. To support the supply of glucose, the glucose polymer *glycogen* serves as an energy reserve. The other sugar that is prominent in the human diet is *fructose*, named for the Latin **fructus** 'fruit', a *ketohexose* and a component of the *disaccharide sucrose*, named for **sucré**, 'sugar' in French. The keto- prefix indicates the presence of a keto or C=O, carbonyl group, in compounds known as ketones, with the general structure $R_1R_2C=O$, where neither R1 or R2 are hydrogens, as in fructose. Both terms, keto- and ketone, derive from *acetone*, the name adopted by French chemists in 1839 for what was previously termed *pyroacetic spirit*, while the German chemist Gmelin introduced the general term *ketone* in 1848.

As the name indicates, fructose is a common sugar in fruit and was already part of the diet of our primate ancestors. The problem with fructose is that human metabolism copes poorly with the amount of fructose present in many modern diets. An obsolete name for fructose is *laevulose*, from the Latin **laevus** 'to the left', based on the early observation that a solution of fructose rotated the plane of polarized light to the left, i.e. it is *levorotatory*. Similarly, glucose was *dextrose*, or *dextrorotatory*, from **dexter, dextr-** 'to the right'. The *-ulose* suffix in laevulose has been retained to indicate a ketohexose. Thus, there occur in the pentose phosphate pathway and in the Calvin cycle the phosphorylated derivatives of the ketopentoses *ribulose* and *xylulose* as well as *sedoheptulose-7-phosphate*. Aldopentoses in these pathways are the pentose *ribose*-5-phosphate and the tetrose *erythrose*-4-phosphate.[26]

Pairs of isomers that are not mirror images are known as *diastereoisomers*. When the difference involves just a single chiral centre, then these are said to be *epimers*, where the Greek prefix can be understood to mean 'near to' or 'similar'. A commonly encountered epimer of glucose is the C-4 epimer *galactose*, which takes its name from the Greek word for milk, **gala, galakt-**, and occurs in the milk disaccharide *lactose*, which is named after the Latin for milk, **lac, lactis**—see also entry for **MILK**. Galactose also occurs as part of mixed *oligosaccharides* covalently bonded to amino acid side chains of **GLYCOPROTEINS**. *Mannose* is the C-2 epimer of glucose and was originally isolated as its reduced derivative *mannitol* from a tree known as the manna tree, named after the manna from heaven mentioned in the Bible, **manna** meaning 'gift' in Hebrew. Mannose also occurs in glycoproteins. The addition in the cellular *Golgi complex* of mannose-6-phosphate to a set of newly synthesized hydrolytic enzymes, the acid hydrolases, serves to target them to the *lysosome*. Polysaccharides consisting mainly of mannose residues, mannans, commonly occur in plants, either as cell wall material or as a carbohydrate reserve. Glucose and the other aldohexoses exist in solution as six-membered rings as a result of a reaction between the C-1 aldehyde and the OH group at C-5 to form *glucopyranose*, named for the structural similarity to the compound *pyran*. A consequence is the presence of an additional *chiral centre* at C-1. The two resulting isomers, α- and β-D-glucopyranose, are known as *anomers* and C-1 as the *anomeric* carbon. Fructose can form both six-membered and five-membered rings, the latter referred to as *fructofuranose*. Furanose is based on its structural similarity to *furan*, originally known as *furfuran*, which derives from the Latin **furfur** meaning 'bran'.

Pentose sugars such as ribose and deoxyribose also form *furanose* rings as occur in DNA and RNA. In the case of the free sugars, an equilibrium exists between the ring and open-chain forms, the latter usually present in very small amounts. This equilibrium means that ring formation of aldose sugars does not abolish their ability to act as reducing agents.

Two other sugars that occur in glycoprotein oligosaccharides (see below) are *fucose*, which corresponds to *6-deoxygalactose*, the name derived from the Greek **phykos** for 'seaweed', and *sialic acid*, a nine-carbon sugar with a carboxyl group, named for its isolation from saliva (Gk. **sialon**). Another name for sialic acid is *N-acetyl neuraminic acid*, which was detected as a component of a *glycolipid* in the brain. For the structures of monosaccharides go to https://en.wikipedia.org/wiki/monosaccarides.

DISACCHARIDES. Common disaccharides are *maltose, lactose, sucrose,* and *trehalose*. Maltose consists of two glucose residues and was first isolated from brewing *malt* and is the result of the incomplete breakdown of starch. Lactose is present in the milk of nearly all mammals and consists of glucose joined to galactose. Sucrose is common table sugar obtained from sugar cane or sugar beet and consists of fructose and glucose. Trehalose, like maltose, consists of two glucose residues but differently bonded. For disaccharides, polysaccharides, etc., we need to specify which carbons participate in what is termed the *O-glycosidic bond* that joins the constituent monosaccharides. If the bond involves anomeric carbon atoms, then these may be bonded in either their α- or β- configurations. Thus, the complete descriptions of these four disaccharides are as follows:

Maltose: α-D-glucopyranosyl-(1 → 4)-α-D-glucopyranose
Lactose: β-D-galactopyranosyl-(1 → 4)-α-D-glucopyranose
Sucrose: α-D-glucopyranosyl-(1 → 2)-β-D-fructofuranose
Trehalose: α-D-glucopyranosyl-(1 → 1)-α-D-glucopyranose.

It is apparent that trehalose and sucrose are not reducing sugars since in both cases, the C-1 carbon of glucose is bonded to the other sugar. Both maltose and lactose have a glucose residue, which is able to assume an open-chain form containing an aldehyde group i.e. a reducing agent.

Trehalose was named by the French chemist Berthelot, who isolated it in 1859 from *trehala manna*, also known as Persian manna, a waxy secretion synthesized by weevil larvae for their cocoons. Trehalose occurrence is widespread in bacteria, fungi, plants, and invertebrates. Originally regarded as simply an energy store, it is now known to be synthesized in response to stress such as heat and desiccation and appears to prevent cell damage under these conditions by forming a gel. Organisms able to withstand long periods of drought do so by entering a state of suspended animation known as *anhydrobiosis*. A characteristic feature of anhydrobiotic organisms is their synthesis of high concentrations of non-reducing sugars such as trehalose, during the induction of anhydrobiosis. Sucrose performs a similar function in other plants. Trehalose has also been found useful for *cryopreservation* of cells—see under **CRYO**- entry.

POLYSACCHARIDES. Sugar residues also form long polymers, i.e. polysaccharides, which may be branched or cross-linked. These can serve as energy reserves such as the starch produced by wheat, potatoes, and rice on which most of the world's population depends as a source of carbohydrate, or perform a structural function such as that fulfilled by cellulose in plants.

Glycogen, the reserve carbohydrate of animals, can be rapidly mobilized by *glycogenolysis*. It consists of chains of glucose residues joined by α-1,4 linkages. It is a highly branched molecule because of the occurrence of *α-1,6 linkages* approximately every ten glucose residues. After a meal, when glucose is in excess, glycogen synthesis occurs and is degraded when glucose is required for ATP synthesis. As would be expected, in order to maintain a constant blood glucose level, these pathways are highly regulated.

The reserve polysaccharide synthesized by plants is *starch*, which consists of long chains of glucose residues joined by α-1,4 linkages known as *amylose*, from the Latin for starch, **amylum**. A second variety of starch, called *amylopectin* (**pektos** Gk. 'congealed'), is similar but contains branches due to *α-1,6 linkages* approximately once every thirty glucose residues. In most plants, the ratio of amylose to amylopectin is around 70:30. Enzymes that degrade starch are *amylases*. See also entry for DIET and EVOLUTION, under **MILK**.

Glycogen and starch, in spite of their high molecular weights, form relatively compact structures. This is partly due to their branched structures but also because linkage of glucose residues by α-1,4 bonds, as in starch, results in bends between successive monomers and the formation overall of an open *helical structure*. This can accommodate *guest molecules* such as *iodine*, which provides a simple colour test for the presence of starch. A completely different stereochemistry results when glucose residues are joined by β-1,4 glycosidic linkages, which occur in *cellulose* and which generate a *linear* structural polymer with high tensile strength. In wood and other plant fibrous materials, cellulose molecules align with each other and are further stabilized by hydrogen bonds between neighbouring molecules. Another structural polysaccharide is *chitin*, derived from the Greek **khiton** meaning 'the garment worn next to the skin or a tunic'. Chitin is a tough protective polysaccharide that consists of cross-linked *N-acetylglucosamine* units joined in β-1,4 linkages and forms the *exoskeleton* of **arthropods** and is found in the cell walls of *fungi*. See also **GLYCOPROTEINS**. For more information on carbohydrates go to: https://en.wikipedia.org/wiki/carbohydrates,

CARBON: **carbo, carbonis** (L.) and Gk. **anthrax, anthrak**- 'live coals, charcoal'. The diminutive of **carbo** is **carbunculus**, which had a number of meanings, including 'a small burning coal', 'a precious stone', or 'a *carbuncle* produced by disease such as anthrax'. *Anthracite* is a particularly pure form of coal; anthracene, $C_{14}H_{10}$, is a tricyclic aromatic hydrocarbon obtained from coal. Organic chemistry = carbon chemistry.

CARTILAGE: **cartilago** (L.), **khondros** in Greek. Cartilage is a type of *connective tissue* consisting of a strong but flexible matrix composed of the proteins *collagen*, *elastin*, and *proteoglycans*. Proteoglycans consist of large proteins, to which are attached *glycosaminoglycan* groups such as *chondroitin sulphate*—see also PROTEOGLYCAN heading under **GLYCOPROTEINS**. Different cartilages are distinguished on the basis of composition and function: *hyaline* cartilage, **hyalinos** (Gk.) 'glassy', is found in joints, such as knee joints, where it provides a cushioning effect, *elastic* cartilage as in the outer ear, and *fibrocartilage* found in intervertebral discs. Proteoglycans are highly negatively charged and therefore bind large amounts of water. The cushioning effect of joint cartilage is due to the reversible squeezing out of some of this water.

Cartilage is produced by CHONDROCYTES that become embedded in the cartilage matrix. Unlike other tissues, cartilage has no blood supply and the chondrocytes rely on diffusion of oxygen and nutrients and this is aided by the compression and re-expansion that accompanies the cushioning effect. Cartilage provides the early skeletal framework, which is progressively replaced by bone mineral during infancy and adolescence in most higher organisms. Exceptions are the CHONDRICHTHYES: **khondros + ikthus** 'fish', the cartilaginous fish such as sharks, whose skeletons are permanently composed of cartilage; cf. OSTEICHTHYES: **osteon** (Gk.) 'bone', the bony fish.

ACHONDROPLASIA: **a-** 'without' + **khondros + plasis** 'development', meaning literally 'absence of cartilage formation'. This mutation is responsible for what is the most common form of *dwarfism* in humans where the limbs are disproportionately shorter than the trunk. It is an *autosomal dominant* disorder resulting in overexpression of one of the forms of a regulatory protein known as *fibroblast growth factor receptor 3* (FGFR3). A more severe mutation in this protein gives rise to *thanatophoric (death-bearing) dysplasia* (TD), which as the name suggests is usually lethal, **thanatos** (Gk.) 'death', cf. *euthanasia*. Normally this protein acts as an inhibitor of long bone growth so that its overexpression results in abnormally short bones. Mice deficient in FGFR3 show skeletal overgrowth. CHONDRODYSPLASIA is a term that sounds like a synonym for achondroplasia; however, it is used as a general heading for a diverse group of diseases related to *bone formation*. A 2009 paper describes the situation in some nineteen breeds of dogs such as corgis and dachshunds that accounts for their relatively short legs and which is referred to as a chondrodysplasia. Although the phenotype is broadly similar to achondroplasia, the mutation here causes overexpression of *fibroblast growth factor 4* (FGF4), i.e. the growth factor itself as distinct from one of its receptors. The mutation involves *retrotranscription* of the FGF4 mRNA which is unusual and must have occurred in the common ancestor of these breeds.[27] See also entry for RETROTRANSPOSONS under **GENOME** heading.

HYPOCHONDRIA: **hypo-** 'below' + **khondros**, the current meaning being a *phobia* or unnecessary anxiety about one's health. Tom Stoppard, the English playwright, is quoted as saying, 'My wife, the doctor, says I am not really a hypochondriac—I just think I am.' *Hypochondria* has undergone a relatively recent change of meaning. Hypochondria, meaning *depression*, was used interchangeably with *melancholy* (**melanos** + **cholos**, meaning 'black bile') which goes as far back as Hippocrates, fifth and fourth century BCE. Galen (129–200 CE) thought melancholia was due to a superfluity of black bile on the brain. The term *hypochondria* arose because the ancients associated feelings of anxiety and melancholia with the **hypochondrium**, the part of the body extending from the cartilage associated with the ribs down to the groin. It was not till 1904 that it was recommended that the term *melancholy* be replaced by depression. See also entry for **BILE**. The *chondrocyte*, the cell type that synthesizes cartilage which precedes most bone growth, is thought to have been central to evolution of the vertebrate skeleton.

CENTRIFUGE: **centrum** (L.) 'centre of a circle' + **fugare** 'to cause to flee' (as in the related words *fugitive, refuge, refugee*). A centrifuge is a piece of equipment designed to subject solutions of macromolecules or suspensions of particles to *high g values* by spinning in a *rotor*, in order to separate, for example, the components of a cell extract on the basis of their size or density, or to collect fine precipitates. A wide variety of centrifugation protocols have been developed for application to particular situations. During centrifugation, the *centrifugal* force acting on the rotor is balanced by the *centripetal* force (**petere**, 'to seek') acting in the opposite direction. It is best not to be standing beside the centrifuge if these two forces suddenly become unequal. One application of the centrifuge uses the *analytical* centrifuge, in which the progress of sedimentation can be followed photographically to determine a *sedimentation coefficient*; *S*. *S* depends on size but also on shape of the particles being centrifuged. The analytical technique is now not often used, although large complexes such as *ribosomes* and their *subunits* are still referred to by their sedimentation coefficients as 80S, 60S, 40S, etc.

CEPHALO-: Greek root meaning 'head or skull'. The *Cephalopoda* are the class of molluscs comprising octopus, squids, and cuttlefish while the *Cephalochordata* are a small group of marine invertebrates, *lancelets* consisting of species of *amphioxus*, described as the closest living relative of vertebrates, also as 'like a vertebrate with everything subtracted'. The *amphioxus* genus is *Branchiostoma*: **brankhia** pl. 'gills' + **stoma** 'mouth', indicating it is a filter feeder. CEPHALOSPORINS are semisynthetic antibiotics produced by the fungus genus *Cephalosporium*. ENCEPHALO-: **enkephalos** 'relating to the brain', literally 'inside the head', hence *encephalitis*, an *encephalopathy*, as in *bovine spongiform encephalitis* or mad cow disease. *Encephalization* is a term for the increasing size of the *hominin* brain over evolutionary periods of time. ENCEPHALINS are a pair of pentapeptides with sequences Tyr Gly Gly Phe Met and Tyr Gly Gly Phe Leu which act as *ligands* for the δ-opioid receptor and thus serve as endogenous *analgesics*. The encephalin precursor is a protein of 267 amino acids containing six interspersed Met-encephalin sequences and one Leu-enkephalin. The Latin word for 'head' is **caput**, which gave rise to the verb **praecipitare**, meaning 'to throw down headlong' and so to *precipitate* in English. *Occipital*: **ob-** 'against' + **caput**: relating to the back of the skull or head. In Latin, the word for 'brain' was **cerebrum**. In English, this refers to the major portion of the brain including the *cerebral hemispheres*, enclosed in the skull or *cranium* (**kranion** Gk.). The diminutive **cerebellum** or 'little brain' is situated below and behind the cerebrum.

CHELATION: khele (Gk.) 'claw' is where a central atom or ion, typically a *metal ion*, forms a stable complex with a chelating agent whose structure is such that it is able to form multiple *coordinate bonds* with the central atom. A commonly employed chelating agent is ethylenediaminetetraacetic acid (EDTA), which strongly binds ions such as Mg^{++} and Ca^{++}. Naturally occurring examples include haem, chlorophyll, and numerous metalloenzymes.

CHEMISTRY: it was not until the seventeenth and eighteenth centuries that the beginnings of modern chemistry began to emerge. New concepts and a new language were badly needed. Robert Boyle (1627–1691) is sometimes cited as the founder of chemistry. Certainly, he adopted the experimental approach, built the apparatus he needed to determine Boyle's law as well as making other contributions to chemistry and was one of the founders of the Royal Society in 1660. However, it was not until the second half of the eighteenth century that significant further progress in chemistry took place. One reason for this delay was the proposal by Stahl in 1705 of the *phlogiston theory*, which said that combustion consisted of the liberation of the hypothetical material phlogiston (**phlogizein** Gk. 'to burn'). Another reason for the lack of progress was the continuing influence of alchemy. Isaac Newton (1642–1726), having revolutionized physics, also devoted much time to studying alchemy. In one sense, alchemy led to the development of modern chemistry, but the superstitions and misconceptions of alchemy had to be discarded before progress was possible.

ALCHEMY comes from the Arabic **al-kmiya**. Its earlier etymology is disputed. One suggestion is that it derives from the Greek **khymeia** or **khemeia**, originally meaning the art of fusing metals, which may have led to the idea of transmuting base metals such as lead into silver or gold. Another suggestion is that it derives from the Egyptian **khemi**, describing the rich black soil along the banks of the River Nile. In any case, it has a long history stretching back to the ancient Greeks and Egyptians. Early alchemy was much preoccupied with a search for the philosopher's stone, which at various times was reputed to be able to turn a base metal such as lead into silver or gold, cure disease, or confer immortality. As the centuries went on, there developed a huge alchemical literature renowned for its obscurity and unreliability, plagued by inconsistent terminology. All the same, alchemy contributed to the development of ore refining, metalworking, the manufacture of gunpowder, ink, dyes, paints, cosmetics, ceramics, and glass, and leather tanning. These practical advances all occurred within an alchemical outlook that was mystical, spiritual, and occult. Alchemy continued to flourish until towards the end of the eighteenth century when people such as Lavoisier and his contemporaries in several European countries began to distinguish the chemical wood from the alchemical trees. Where alchemy did contribute to chemistry was in laboratory techniques such as distillation and ways to prepare various chemicals.

When Joseph Priestly (1733–1804) isolated oxygen in 1774, he called it 'dephlogisticated air'. Cavendish (1731–1810) reported in a paper on 'factious airs' in 1766 that he had obtained a light gas by treating metals with acid. This gas burned readily in air, and he thought that it might be pure phlogiston. According to the received wisdom then current, phlogiston was given off when metals were heated and converted to calxes. When it was pointed out that heating metals in air actually led to a weight gain, the phlogiston proponents responded that evidently phlogiston can sometimes have negative weight. It was the enduring contribution of Lavoisier (1743–1794) to realize that when metals were heated, they combined with Priestley's dephlogisticated air, which he renamed *oxygen*. Lavoisier made two major contributions: first, he put the long-held phlogiston theory out of its misery, while with his colleagues de Morveau, Berthollet, and Fourcroy, he set about reforming chemical nomenclature. This involved renaming the elements as they were then known. The products obtained by heating metals were renamed metal oxides. In the meantime, it became clear that the result of combusting Cavendish's gas with that of Priestley was the formation of water, so the former gas was named *hydrogen*: from **hudros** + **-gen** 'water former'. The fraction of air unable to support life became *azote* from **a** + **zoe** (Gk.) 'without life' and still the word used for nitrogen in France, while *nitrogen*, 'nitre-forming', nitre being KNO_3, was the word adopted in Britain. The new nomenclature was written up by Lavoisier in *Méthode de nomenclature chimique* (1787). He summarized his chemical ideas in *Traité élémentaire de chimie* (1789), described as 'a book that stands in relation to the development of chemistry as Newton's *Principia* does to physics'.[28]

Torbern Bergman (1735–1784) was a Swedish chemist who collaborated with de Morveau in attempts to make chemical nomenclature more systematic. It was Bergman who proposed that the names of metals should end in *-um*, a suggestion followed by Humphry Davy (1778–1829) when he later isolated sodium, potassium, barium, strontium, calcium, and magnesium. Bergman in turn was influenced by his compatriot Linnaeus (1707–1778), who had introduced the *binomial nomenclature* in biology and who, through Bergman, played an indirect role in establishing the binomial nomenclature used in inorganic chemistry today, **binominis** (L.) 'having two names'.[29]

CHIRALITY: **kheir** (Gk.) 'hand'. Chirality is the property of molecules known as *handedness*. The term applies to molecules that can exist in two separate similar but different forms known as *enantiomers*. This is most easily seen with a simple molecule such as the amino acid alanine, where the α-carbon atom has four different *substituents*: H, CH_3, NH_2, and COOH. Here the α-carbon is also referred to as an *asymmetric carbon* or as a *chiral centre*. The four substituents will be in a *tetrahedral* arrangement in space, and there are two possible arrangements, designated as D or L. With the exception of glycine, all amino acids that occur in proteins have the L configuration about the α-carbon atom. Glycine with two hydrogen atoms attached to the α-carbon is *achiral*. Some D amino acids are found in bacterial cell walls and in some antibiotics. The labels D and L indicate configurations that are the same as those for a reference molecule; for amino acids, this is *glyceraldehyde*.

Another definition of a chiral molecule is that it is not able to be *superimposed* on its mirror image. This is where the idea of *handedness* comes in. Your right hand is the *mirror image* of your left hand, and one is not superimposable on the other. Enantiomers are in general not distinguishable by most chemical tests, but enzymes, whose active sites have precise three-dimensional conformations certainly can, while a physical test that does discriminate is their interaction with *polarized light*. The enantiomer that rotates the plane of polarized light clockwise is termed *dextrorotatory* and the other *laevorotatory*, **dexter** and **laevus** in Latin meaning right and left respectively. Hence, enantiomers are sometimes referred to as *optical isomers*. The naturally occurring form of glucose is dextrorotatory and is still occasionally referred to as *dextrose*, and similarly, fructose was *laevulose*. The direction of rotation of polarized light, designated *d* or *l*, does not predict the absolute configuration of an isomer, i.e. whether it is D or L.

A mixture of equal quantities of both enantiomers is known as a *racemic mixture*, from **racemus** (L.) 'bunch of grapes', originally associated with a racemic mixture of *tartaric acids* obtained from grape juice. In 1847, Pasteur (1822–1895) noticed that crystals of ammonium sodium tartrate occurred in two forms. He proceeded to separate the two forms manually and showed that they corresponded to dextrorotatory and levorotatory forms. This was an important step in the development of ideas about chemical structure.

CHIROPODIST: **khiro-** + **pous, pod-** 'foot'. In spite of the word containing stems for both hands and feet, chiropodists confine themselves to the treatment of corns and bunions. *OED2* suggests the word derives from the Greek **chiropos** meaning 'having chapped feet'.

CHIROPRACTOR: **khiro-** + **praktikos** 'active, effective': one concerned with treatment of diseases and complaints by manipulation of the body, especially the spinal column.

CHIROPTERA: **khiro-** + **pteros** 'wing': order of mammals, the bats. Here the finger bones are elongated and covered with membranes adapted for flight. Bats are a reservoir for diseases transmissible to humans—see entry for **ZOONOSES**.

CHIRURGIA (L.) 'surgery', from the Greek **kheirourgia** 'handiwork or surgery', from **kheir** + **ergon** 'work'. The fourteenth-century French physician Guy de Chauliac was unusual in being surgically trained, and was preoccupied with the issue of trust. In his seven-volume *Chirurgia Magna*, he wrote that 'a doctor should be willing to learn, be sober and modest, charming, hard-working, and intelligent. He should care for the rich and poor alike for medicine is required by all. If payment is offered, he should accept it; but if it isn't offered, he shouldn't demand it'.

CHOLESTEROL: **khole** (Gk.) 'bile' + **stereos** 'solid' is one of the main types of lipid in membranes, being a component of all eukaryotic plasma membranes. It is also a precursor of bile salts and of several types of *steroid hormones*. *Sterol* was the term introduced in the early days of chemistry to describe compounds containing hydroxyl groups but which, unlike the low-molecular-weight compounds methanol, ethanol, etc., were insoluble, i.e. *solid* rather than liquid. The current definition of a **sterol** is any steroid that contains one or more hydroxyl groups and a hydrocarbon side-chain, such as chole*sterol*, lano*sterol*, and many others, while a **steroid** is any of a large group of compounds that share the four-fused-ring system found in cholesterol, i.e. *cyclopentanoperhydrophenanthrene*. This basic ring system with no substituents, side chains, or double bonds has been christened *gonane* since the steroid sex hormones produced by the *gonads*, Greek: **gone** 'seed', lack the hydrocarbon side chain and can be regarded as derivatives of gonane. The *stereo-* combining form has expanded from meaning simply 'solid', to include '3D shape' as in *stereoisomer*. For the structure of cholesterol, go to https://en.wikipedia.org/wiki/Cholesterol. Likewise, structures for cholesterol derivatives can be readily found by searching for gonane, androgen, oestrogen, etc.

CHOLESTEROL SYNTHESIS. While the four-ring system of cholesterol looks like it might be something that came out of a coal mine, in fact eukaryotic organisms manage to put it together starting with three molecules of *acetyl CoA*. The numerous steps in its synthesis can be summarized as follows:

$3 \times 2C \rightarrow 6C \rightarrow \rightarrow \rightarrow 5C + CO_2$. 5C is isopentenyl pyrophosphate, an activated *isoprene* unit. $3 \times 5C \rightarrow 15C$ farnesyl pyrophosphate. $2 \times 15C \rightarrow 30C$ *squalene*. Squalene takes its name from the *Squalidae*, a family of sharks that traditionally has been prized for their shark liver oil, which contains squalene. Activation of squalene results in a concerted electron rearrangement and formation of the steroid ring system. The initial product is C30 *lanosterol*. Lanosterol is the precursor of all animal and fungal steroids but not those of plants. It was first isolated from wool (**lana** L. 'wool'), where it is a component of *lanolin*, or wool grease, that serves a waterproofing function for sheep. It is the equivalent of **sebum** (L. 'tallow, grease') secreted by *sebaceous glands* in other mammals that serves to lubricate skin and hair. A further series of reactions converts lanosterol to C27 cholesterol. The equivalent of cholesterol in fungal membranes is *ergosterol*, named for the *ergot* fungus from which it was first isolated.

The 6C intermediate in the scheme above is *hydroxymethylglutaryl CoA* (HMG CoA). It is the substrate for HMG CoA reductase, which is the *committed* step in cholesterol synthesis. Regulation occurs here at both the *transcriptional* and *translational* levels. HMG CoA is mainly synthesized in the liver, some also being made in the intestine, while the rate of synthesis varies with dietary intake. This enzyme is the target for the group of drugs known as *statins*. See also under ENZYME INHIBITION under **ENZYMES.**

CHOLESTEROL DERIVATIVES

BILE SALTS: Half of the body's cholesterol is converted to bile salts. These are synthesized in the *liver*, stored in the *gall bladder*, and released into the *intestine*. There, they solubilize dietary lipids to help their absorption in the gut, along with the *fat-soluble vitamins*. Most bile salts are reabsorbed from the intestine in a process known as *enterohepatic recirculation*. Bile salts in humans are derivatives of cholic acid, a C24 derivative of cholesterol with a side chain terminating in a carboxyl group. The main bile salts are, first, *glycocholic acid*, formed by an amide linkage between the C24 carboxyl and the amino group of *glycine* (H_2N-CH_2-COOH). A second involves a similar reaction with *taurine* (H_2N-CH_2-CH_2-SO_3H) to form *taurocholic acid*, first isolated from ox bile, (**tauros** Gk. 'bull').

See under **VITAMINS** for VITAMIN D, which is also derived from cholesterol and for CORTICOSTEROIDS under **ADRENAL GLANDS.**

SEX HORMONES: **horme** (Gk.): 'impulse'. These are the *androgens* (*androstenedione* and *testosterone*) and the estrogens (*estrone* and *estradiol*). The initial step in their synthesis from C27 cholesterol is truncation of the side chain to C21 pregnenolone, which is also the precursor for the C21 corticosteroids—see **ADRENAL GLANDS.** In both cases, pregnenolone is converted to C21 progesterone. Progesterone is then hydroxylated at position 17 followed by removal of the 2C side chain from position 17 to give androstenedione. Conversion of the resulting 17-keto group to 17-OH gives testosterone. These two androgens may then be converted to oestrone and oestradiol respectively by conversion of steroid ring A to a fully aromatic ring, which necessarily involves loss of the angular methyl group at position 7. The enzyme involved here is *aromatase*. *Progesterone* derives from **gestare** (L.) 'to carry in the womb', cf. *gestation*, androgen is from **andro-** (Gk.) 'man', while estrogen derives from **oistros** 'gadfly'—see entry for **OESTRUS.** For sex hormone structures see https://en.wikipedia.org/wiki/androgen. Likewise for estrogen.

ECDYSONE: **ekduein** (Gk.) 'to take off'. *Ecdysis* is the process whereby reptiles such as snakes shed their skin and insects their outer *cuticle*. Molecular phylogeny studies have linked *arthropods* to another group of *ecdysing* animals that includes *nematodes* to form the group *Ecdysozoa*. *Ecdysone* is a *steroid hormone* that regulates moulting in arthropods. H. L. Mencken invented the word *ecdysiast* to describe a striptease artist.

CHROMATOGRAPHY: **khroma, khromat-** 'colour' + **graphia** 'writing'. This covers a variety of techniques for separating mixtures of molecules. The mixture is dissolved in a fluid termed the *mobile phase* and passed through a medium termed the *stationary phase*. Separation of the components of the mixture depends on their relative affinities for the mobile and stationary phases. The name arose because the technique was first applied to mixtures of naturally coloured compounds. AFFINITY CHROMATOGRAPHY consists of the use of a stationary phase modified by attachment to it of a compound for which a particular component of a mixture has specific affinity.

CHROMOPHORE: **khroma** + **-phoros** 'bearing', literally colour-bearing, so designates the chemical group that confers colour, such as the haem group in haemoglobin. CHROMIUM from **khroma** (Gk.) meaning 'colour', because many of its compounds are coloured. See also entry for **ABSORB**.

CHRONIC: from **khronikos** 'of or concerning time' from **khronos** 'time': describing a persistent or long-lasting illness, as in 'chronic fatigue syndrome'. In the context of disease, *chronic* is often contrasted with *acute*.

CHRONICLE (*n*. and *v*.): the continuous register of events in order of time, i.e. in *chronological* order. SYNCHRONOUS: happening at the same time, simultaneous. CHRONOBIOLOGY: the scientific study of *temporal**
or periodic phenomena in biology—see **CIRCADIAN RHYTHM** below. Another example of chronobiology is HETEROCHRONY, a concept introduced by the German zoologist Ernest Haeckel in 1875. The idea is that the varying sizes and shapes of different animal species can be accounted for by different extents of developmental growth in different tissues. Thus, giraffes gained their long necks not by having additional vertebrae but by extended development of the usual number of neck vertebrae. This implies the presence of tissue-specific oscillating mechanisms or developmental clocks, the details of which are beginning to be understood.[30]

*From **tempus, tempor-** (L.): 'time', hence *temporal*: used in various contexts including to mean 'pertaining to time'; also, *temporary*: 'transient, lasting for a limited time'.

CHYME: **khymos** (Gk.) 'juice'. Chyme is the semifluid mass of partly digested food expelled by the stomach into the duodenum. Chyme, containing HCl secreted by the stomach, has a pH of about 2. To raise the pH towards neutrality, the *duodenum* secretes first a hormone *cholecystokinin*, which causes the gall bladder to release alkaline bile into the duodenum, and second, the hormone *secretin* to stimulate secretion by the *pancreas* of large amounts of sodium bicarbonate to raise the pH of chyme to pH 7, i.e. neutrality.

CHYMOSIN is a proteolytic enzyme obtained from the *abomasum*, the fourth stomach of ruminants and is the enzyme used to coagulate milk during the first step in cheesemaking. It is an enzyme similar to *pepsin* with general proteolytic activity at low pH. At pH 6.7, the pH of milk, however, its activity is restricted to cleavage of a single peptide bond in kappa-casein, a single component of the casein complex. This results in the formation of a casein curd with subsequent expulsion of whey.

CHYMOTRYPSIN: **khymos** + **tripsis** 'friction' from **tribein** (Gk.) 'to rub', this being an early method of extracting chymotrypsin from the pancreas. Chymotrypsin is a mammalian proteolytic enzyme synthesized in the pancreas as the inactive precursor *chymotrypsinogen*. It is activated by *trypsin*, a related proteolytic enzyme, and functions as a digestive enzyme in the small intestine. The active enzyme preferentially cleaves peptide bonds on the carboxyl side of the aromatic side chains of *tyrosine*, *tryptophan*, and *phenylalanine* and large hydrophobic residues such as *methionine*.

MESENCHYME: **mesos** 'middle' or 'intermediate' + **enkhuma** (Gk.) 'infusion'. Mesenchyme is described as a loosely organized tissue, chiefly *mesodermal* in origin and, in vertebrates, the embryonic tissue that develops into *connective* and *skeletal tissues*, including blood, lymph, and muscles. There is evidence that, at least in some situations, the *metastasis* of cancer cells, i.e. their spread to other organs, results from a transition of epithelial cells to mesenchymal cells. PARENCHYME refers to the specialized tissue of an organ as distinct from its *connective tissue*.

CHYLE: **khulos** (Gk.) 'juice' via Latin **chylus**. A milky fluid formed in the small intestine by the action of the pancreatic juice and bile on the *chyme*, and contained in the *lymphatics* of the intestine, also called *lacteals*.

CHYLOMICRON: **khulos** (Gk.) + **mikron** 'small'. Chylomicrons are small lipoprotein particles that function to transport dietary lipids, with a size range of 75 to 200 nm, and their composition is likewise variable. They consist mainly of an internal core of triglycerides and are stabilized by an external *amphipathic apolipoprotein* and the polar *head groups* of *phospholipids*. They also transport fat-soluble vitamins and cholesterol. After a meal, chylomicrons appear initially in the lymph system then the blood. Chylomicrons are generated by absorptive cells of the small intestine known as *enterocytes*.

CIRCADIAN RHYTHM: **circa** 'around' + **dies** 'day' + **rhythmus** (L.) 'regular recurring motion'. Most biological processes display a built-in rhythm that oscillates every twenty-four hours. These are also described as *diurnal rhythms*, from **diurnus** (L.) 'daily'. Jet lag after long flights that cross multiple time zones is a manifestation of the disruption of this rhythm, while shift workers are liable to suffer adverse health outcomes due to chronic *circadian misalignment*. Circadian behaviour has been observed in animals, plants, fungi, and bacteria and occurs in nearly all types of cells. These rhythms are driven by a *circadian clock* which is *entrainable*, that is to say it is adjustable in response to external cues, and the cue here is daylight. The clock period, twenty-four hours, obviously corresponds to day length, and the ubiquity of circadian clocks shows they are an ancient invention that provides an adaptive advantage.

The basic mechanism of the circadian clock consists of a transcriptional-translational *oscillator* or *feedback loop* in which the core components cycle with a twenty-four-hour oscillation period. These core components, amounting to thirty different proteins according to one estimate, interact with many other proteins, mainly *transcription factors* which in turn regulate the expression of hundreds, possibly thousands, of other genes, the CCGs, the *clock-controlled genes*. For mammals, a master clock regulator is located in an area of the *anterior hypothalamus*, the *suprachiasmatic nucleus* (SCN). This is the only part of the circadian system that has *retinal innervation*, which enables entrainment by the sun.

It came as a surprise that clock genes are expressed in nearly all cells, their diurnal behaviour being synchronized with that of the SCN master clock in the brain. Ablation of the SCN abolishes circadian rhythm. This work has drawn attention to the importance of adequate light exposure. The average health of patients in nursing homes improved when their exposure to light was increased. The SCN clock adjusts to day length, controls the sleep/wake cycle, and influences mood and cognitive ability. Peripheral clocks influence metabolic pathways to the extent that frequent eating of energy-dense foods appears to cause circadian misalignment and predispose to metabolic disease. Good health depends on when you eat as well as what you eat. Three of the researchers who pioneered work on circadian rhythm were awarded the 2017 Nobel Prize in Physiology or Medicine.[31]

CIRCLE: from **circus** (L.) and the Greek **kirkos**, both meaning 'ring' or 'circle'. Circus Maximus was an arena in Rome estimated to accommodate 100,000 people which was used for chariot races and other public games.

Circa was a preposition meaning 'near' or 'in the vicinity of'. *Circa* has come into English but indicates an approximate time, e.g. *circa* 1850, rather than place. It is usually abbreviated to *ca.* or *c.*

CIRCULUS: the diminutive of **circus**, also meaning circle, a circular object, or a circle of friends in Latin and giving rise in English to *circular*, *circulate*, and the *circulatory* (i.e. the blood and lymph) systems.

CIRCUM-: see chapter 3.

COLLOID: kolla (Gk.) 'glue', cf. *collagen*. Many biological molecules such as DNA, polysaccharides, and many proteins are very large molecules. Traditionally, these have been referred to as colloids or as having colloidal properties. The same applies to large structures containing multiple components such as *ribosomes* and *viruses*. One of the properties of colloids is their ability to *scatter light* which occurs as the size of the particles approaches the wavelength of light. For this reason, concentrated suspensions of, for example, virus particles are visibly *opalescent*. Formally, a colloidal suspension consists of two phases—the *solvent phase* and the *disperse* or *solute* phase. A colloidal solution is in between a *true solution* and a *suspension*.

COLOURS: color (L.), **khroma** (Gk.).

> Why grass is green, or why blood is red
> Are mysteries which none have reached unto,
> —John Donne (1572–1631)

For an update on the subjects of the above lines, see entries for **PHOTOSYNTHESIS** and **BLOOD**.

The perception and description of colour by the ancients has caused much modern discussion. It is said for instance that the Greeks had no word for 'blue' and that nowhere in the New Testament does the word 'blue' occur. Pliny claimed that the painter in Classical Greece used only four colours: black, white, red, and yellow. Much speculation has centred on Homer's description of the sea as 'wine-dark'. Homer also has sheep the colour of wine, honey is green, and the sky is often described as bronze. What was Homer smoking? Xenophanes (c. 570–c. 475 BC) described the rainbow as having only three colours: **porphura** (dark purple), **khloros** (green), and **eruthros** (red). Ball (2002) sensibly makes the point that *we* think of colours in terms of the modern spectrum of Newton, namely red, orange, yellow, green, blue, indigo, and violet. Also in chapter 2, he summarizes the explanation of how colour arises.[32] The *spectrum* that the ancients saw was light at one end, dark at the other, with a group of colours in between. Red and yellow were obviously towards the lighter end before darker colour(s). Although in the list below, *kuanos* is given as blue, to the Greeks it was probably sometimes dark blue and at other times just a dark colour.

BLACK: **melas-, melan-** (Gk.) as in *melanin, melanocyte, melanoma, melancholy.* **ater, atri-** as in *atrabilious*, the Latin version of melancholy.

WHITE: **albus** (L.) as in albumin, also used to mean colourless, as in the white of an egg.

candidus (L.): 'shiny white', which derives from the verb **candere** 'to shine, glow, be white' as in *incandescent,* while **candescere** means 'to become white'—a whiter shade of pale perhaps. *Candida albicans* is a yeast-like fungus which upon infection can cause *candidiasis* or thrush. **Candidus**

is also the origin of the word *candidate*, it being the custom for Romans standing for office to wear a white toga.

leukos (Gk.): as in the amino acids *leucine* and *isoleucine* and white blood cells or *leukocytes*.

YELLOW: **flavus** (L.) as in *riboflavin* and the coenzymes *flavin adenine dinucleotide* (FAD) and *flavin mononucleotide* (FMN). The pale-yellow colour of flavoproteins such as succinate dehydrogenase is due to the presence of bound FAD or FMN.

luteus (L.) as in **corpus luteum. Luteus** was the word used by Pliny to describe the colour of the yolk of an egg. A plant known as **lutum** was used to dye fabrics yellow.

xanthos (Gk.): as in xanthophylls, the plant pigments that give rise to autumn colours. The Greeks also referred to the colour **kirrhos**, translated as 'tawny', which is the root for *cirrhosis*, as in 'diseased liver'. Dictionary definitions of tawny include 'a shade of brown, tinged with yellow' or 'dull yellowish brown'.

chryso- 'golden', cf. St John Chrysostom: **chryso-** + **stoma, stomata-** 'mouth', hence 'golden mouth', patriarch of Constantinople 398–404 CE. He evidently had a way with words.

GREEN: **khloros** (Gk.) as in *chlorine* as well as the green plant pigment responsible for photosynthesis, *chlorophyll*.

viridis (L.): synonyms including *virens, viridians*, as well as variations such as *sempervirens, viridissima* are common species names in botany. The OF root **verd** shows up in *biliverdin*, a green intermediate in haemoglobin degradation. Similarly, *verdigris* is the patina that forms on copper or brass, consisting of copper carbonate, also *verdant*, which describes a flourishing pasture.

BLUE: **kuanos** 'dark blue' as in **cyanobacteria**, also known as **blue-green algae**. In some invertebrates, arthropods, and molluscs, the oxygen-carrying protein is *haemocyanin* in place of haemoglobin. Haemocyanin, which contains copper ions, assumes a pale-blue colour when it binds oxygen.

The *cyano-* root occurs in the names of a range of chemicals, many of which are not blue. This came from the accidental synthesis in 1704 of a blue pigment known as *Prussian blue*. Later, much later, the formula for Prussian blue was deduced to be $KFe^{3+}[Fe^{2+}(CN^-)_6]$, the result of potassium ferrocyanide reacting with a ferric salt. Treatment of Prussian blue with acid produces HCN, hydrogen cyanide, originally known as *prussic acid*, and the CN^- ion was called *cyanide* from its Prussian blue precursor.[33] Cyanide is extremely toxic because it binds to the cytochrome oxidase component of the electron transport chain, preventing reaction of electrons with oxygen. See entry for **MITOCHONDRIA**. The word *anthocyanin* literally means 'blue flower'; however, most anthocyanins are red. Organic compounds containing the $-C\equiv N$ group are known as nitriles, e.g. CH_3CN *acetonitrile*.

RED: **eruthro-** (Gk.): as in *erythrocytes*.

purros: from which we get *pyrrole, tetrapyrrole, pyrrolidone*, etc.; **purros** is derived from **pur** (Gk.) meaning 'fire', the root for *pyruvic acid, pyrophosphate, pyrolysis*, and *pyromaniac*.

ruber, rubra, rubrum (L.): as in Rubrum Mare (Red Sea), *rubella* (German measles). *Bilirubin*: the next step in the breakdown of haem after biliverdin.

PURPLE: **porphura** in Greek and **purpura** in Latin, from which we get the word *porphyrin*, the planar molecule consisting of four variously modified pyrrole groups, found in haem, chlorophyll, and vitamin B12. *Porphyrias* are a group of diseases that result from assorted abnormalities in haem metabolism. In some porphyrias, haem precursors excreted in the urine make it turn a reddish-brown colour when exposed to sunlight. In classical times, the dye Tyrian purple was highly prized. This dye obtained from sea snails was produced in Asia Minor from at least 1500 BCE. During the Roman Empire, robes dyed with Tyrian purple were worn only by emperors. There is evidence that considerable variation in hue was common, anything from purple to dark red or crimson. Ball (2002) quotes

Pliny that the colour was that of clotted blood. The Romans also quarried a purple hard rock called *porphyry*, which they regarded as semiprecious.

GREY: **polios** (Gk.). A rare example of the use of this root in English is the disease *poliomyelitis*, **muelos**: 'marrow', together with *poliovirus*, its *neurotropic* cause. It appears that the Greeks themselves avoided its use, an exception being **poliothrichos** 'grey-haired' (**thrix** 'hair'), from which we get *trichology*, the study of hair and its diseases. **Poliosis** was Latin for 'grey hair', likewise in medical English.

COMPOSITION and COMPOUND both derive from **componere** (L.) 'to put together', *ppp.* **compositum**, from **com** + **ponere, positum**, 'to put'. *Composition* is a word that is very much at home in chemistry as well as elsewhere. *Compound* bears marks of its sojourn in France, as do *propound* and *expound*, also from the **ponere** stem. Among its various usages are the chemical, as in 'water is a compound of hydrogen and oxygen', the biological, where a compound structure such as a flower consists of several simpler parts, and the linguistic, where compound words such as *chemolithoautotrophic* are formed from multiple stems.

CRYO-: kruos (Gk.) 'frost'.

CRYOBIOLOGY: the biology of cells and organisms at low temperature.

CRYO-ELECTRON MICROSCOPY. Determination of the three-dimensional structures of proteins and *supramolecular* assemblies such as ribosomes has mainly relied on *X-ray diffraction* of crystals. Hence, materials such as membrane proteins which are difficult or unable to be crystallized are not amenable to the X-ray approach. Recent advances in cryo-electron microscopy (cryo-EM) enable protein and other structures to be determined at resolutions that compare with those obtainable by X-ray diffraction. This also has the advantage that only small amounts of material are required. Cryo-EM consists of preparing specimens in a thin layer of vitreous ice by exposure to liquid nitrogen or helium. This reduces the damage caused by exposure to the electron beam and allows direct imaging of the specimen, as distinct from the earlier method involving negative staining with heavy metals. Direct imaging means that very large numbers of individual particles that are randomly oriented can be analysed and their orientation determined *computationally*. An additional advantage of cryo-EM is that snapshots can be obtained of molecules that change their structure as part of their function. A crucial development has been the use of direct detector devices (DDDs) that greatly improve resolution compared to that obtained using earlier methods of recording data. It was the use of DDDs that led to the unexpected discovery that exposure to the electron beam caused specimens, along with the ice, to actually move. When this was corrected for, again computationally, together with improvements in image-processing algorithms, it became possible to obtain 3D structures with resolutions around 3Å. (Å stands for Ångstrom unit, 10^{-10} metre). Cryo-EM is being successfully applied to the determination of numerous complex biological structures. Recently, single-particle protein structures have been reported at atomic resolution. The 2017 Nobel Prize in Chemistry was awarded to a trio of scientists 'for developing cryo-electron microscopy for the high-resolution structure determination of biomolecules in solution'.[34]

CRYOGENICS: the study of matter at very low temperatures as well as the means of achieving those temperatures. This includes the cryogenic preservation of plant seeds in seed banks in attempts to prevent the erosion of plant diversity and involves research to develop new techniques for the preservation of so-called recalcitrant seeds. A commercial application of something similar, known as CRYONICS, consists of the practice of deep-freezing the bodies of people who have died of a currently incurable disease, in the hope that at some time in the distant future they may be able to be thawed and cured. One incurable condition that afflicts people who go in for this would appear to be incurable optimism. It has been suggested, on grounds of economy, that just the head be frozen. Margaret Atwood, the author of novels set in future dystopian worlds: 'Cryonics is a nonstarter: you get your head frozen, the money runs out, your relatives die, and you're cat food.'

CRYOPROTECTANTS, as the name suggests, are compounds that are used to protect cells of different types against the damage that the formation of ice crystals can cause during freezing. Most such protectants are *polyhydroxy* compounds such as *glycerol, fructose*, and *trehalose. Dimethyl sulphoxide* at concentrations around 15 per cent is also widely used. Bacteria, animal cells in tissue culture, and embryos can all be stored long term, usually in liquid nitrogen at -196°C.

CRYSTAL: also derived from the CRYO- root via **krustallos** from **kruos + halos**, literally 'salt ice', used by the Greeks to mean ice or quartz.

CYCLE: **kuklos** (Gk.) 'circle'. The main METABOLIC CYCLES in animals are the citric acid cycle and the urea cycle. In both cases, an initial condensation reaction is followed by a series of steps that ultimately generates one of the components of the initial reaction. In the CITRIC ACID CYCLE, a two-carbon fragment, acetyl CoA, reacts with *oxaloacetic acid* to generate citric acid. Subsequent reactions, including reduction of NAD^+ and FAD, lead to the elimination of two molecules of carbon dioxide and regeneration of oxaloacetic acid. In the UREA CYCLE, *carbamyl phosphate* condenses with ornithine. Several steps later, arginine is hydrolysed by the enzyme arginase to form *urea* and *ornithine*. In the CALVIN CYCLE, the dark reactions of photosynthesis involve carboxylation of the 5-carbon sugar ribulose 1,5-bisphosphate by the carboxylase known as *rubisco* with the formation of two molecules of 3-phosphoglycerate which, by a reversal of the reactions of glycolysis, can form fructose 6-phosphate. Another series of reactions involving the 3-carbon sugars regenerates the 5-carbon substrate for rubisco. The net result is that for every six molecules of CO_2 fixed, one 6-carbon sugar is produced.

The Greek word **kuklos**, quoted above with the meaning 'cycle' or 'circle', can also mean 'a circle of friends'. It was in this sense that in 1865 a small group of Southerners in the US who knew their Greek but were not happy with the result of the recently ended civil war, called itself the Kuklos Clan, which soon became the Ku Klux Klan.

CELL CYCLE or cell division cycle: see entry for **MITOSIS**.

CYCLIC AMP = cyclic 3',5'-adenosine monophosphate. Many hormones trigger the formation of cyclic AMP, which acts as a *second messenger* mediating a physiologic response to the hormone, the *first messenger*.

CYCLOSPORIN: **kuklos + spora** (L.) 'spore'. A cyclic peptide with *immunosuppressive* properties used to prevent rejection of tissue grafts and organ transplants. It was the introduction of cyclosporin that made successful organ transplantation possible. It has now been superseded by other more effective immunosuppressants.

CYTO-: kutos (Gk.) 'vessel', a combining form denoting a cellular component, structure, or process.

CYTOCHROME: **kutos + khroma** 'colour'. A cytochrome is an *electron-transferring* protein that contains a haem *prosthetic group*. A series of cytochromes constitute part of the electron transport chain which channels electrons from NADH to oxygen. This *exergonic* process is coupled to oxidative phosphorylation whereby ADP is phosphorylated to form ATP, an *endergonic* reaction. See entry for **MITOCHONDRIA**. A separate example of a cytochrome is CYTOCHROME P450 (CYP), which is the terminal component of an electron transport chain in adrenal mitochondria and liver microsomes. CYP consists of a large family of haem-containing enzymes termed *monooxygenases*, there being approximately 100 genes in humans that encode CYP enzymes of different specificities. The 450 refers to the wavelength of maximum absorption of the haem group, measured under specified conditions. In the presence of NADPH and molecular oxygen, CYPs insert hydroxyl groups into substrate molecules, examples being numerous reactions involved in the metabolism of eicosanoid and steroid hormones. CYP enzymes also play an important role in the detoxification of foreign substances, i.e. *xenobiotics*. The introduction of hydroxyl groups increases their water solubility and facilitates excretion by providing sites for *conjugation* with highly polar groups such as *sulphate* and *glucuronate* which make the xenobiotics still more water-soluble and excretable.

CYTOPLASM: **kutos + plasma** 'shape'. The cytoplasm of eukaryotic cells is defined as consisting of all cellular components except for the cell membrane and the nucleus. The part of the cytoplasm which does not include organelles such as mitochondria, endoplasmic reticulum, etc. is referred to as the CYTOSKELETON: **kutos + skeletos** 'dried up'. The term *cytoskeleton* encompasses a variety of structures that perform numerous functions. The cytoskeleton helps determine *cell shape, cell movement, cytokinesis,* and the *position* of cell organelles. In eukaryotic cells, the cytoskeleton consists of three main kinds of filament: *microfilaments* (7–8 nm diameter) consisting of linear polymers of *actin* subunits, *intermediate filaments*, and *microtubules*. Intermediate filaments (IF), average diameter 10 nm, are probably the most important determinants of cell shape. The protein components of IF vary depending on cell type and include *keratins* in epithelial cells, *vimentin* (which is widely distributed in different cell types, a family of filament proteins in the axons of neurons), and *lamin*, which is restricted to cell nuclei. A mutation in the lamin precursor is responsible for the disease PROGERIA (**progeros** Gk. 'prematurely old' from **pro** 'before' + **geras** 'old age'). Progeria is a rare disease in children, characterized by premature ageing. Progeria patients have abnormally shaped nuclei. Experiments carried out in mice, using *gene therapy* techniques, have resulted in significant alleviation of progeria symptoms.[35]

Microtubules (diameter 30 nm) consist of a helical array of α- and β-*tubulin* subunits. They are the target of the poison *colchicine*, which binds to tubulin—see under **PLOIDY**. Microfilaments and microtubules, but not IFs, serve as tracks along which other proteins, termed *motor proteins*, move. Actin associates with *myosin* to form the *actomyosin* complex, which is the basis of ATP-driven muscle contraction, while GTP hydrolysis provides the energy for *kinesin* and *dynein* movement along microtubules. Kinesins play a role in protein, mRNA, and vesicle transport and in formation of the spindle during mitosis. Dyneins power the movement of cilia and flagella as well as movement within cells. Sequencing of the human genome revealed the presence of forty different myosins, at least forty kinesins, and about ten dyneins. These numbers reflect the variety of different processes, often tissue-specific, in which these proteins participate, many of which remain to be documented.

DERMA (Gk.) 'skin'. In early embryos of higher animals, there are three tissue layers designated *endoderm*, *mesoderm*, and *ectoderm*. These organisms, which correspond to those showing bilateral symmetry, are referred to as *triploblastic*, while simpler organisms, such as *Cnidaria* and including corals, jellyfish, and hydra, have only two cell layers and are *diploblastic*. Other examples of words with the **derma** root all relate to aspects of the skin: the two main cellular layers of the skin, the *epidermis* and *dermis* protect the body from microbes, pollutants and water loss; also *dermatitis, dermatology, intradermal, epidermal growth factor* (a family of growth factors found in many tissues); *blastoderm*, the outer layer of cells of the *blastula*; *echinoderm*, a member of the class *Echinodermata* such as sea urchins with spiny skin—**echinos** in Greek was the name for both the sea urchin and the hedgehog. *Ostracoderm* is the name of an extinct jawless fish with a heavily armoured body, named from the Greek **ostrakon** meaning 'shell', which they used as a writing surface. At one time in Athens, if your name was written on an **ostrakon**, then you were ostracized, i.e. banished. And if there is a large quadruped in the room with ivory tusks, it is a *pachyderm* (**pachys**, Gk. 'thick'). The practice of preserving dead animals in a lifelike form is *taxidermy*, from **taxis** (Gk.) 'arrangement' + **derma**; cf. *taxonomy*. HYPODERMIC: literally 'under the skin' as in *hypodermic needle*. *Subcutaneous*, from **cutis** (L.) 'skin', has the same meaning, as in *subcutaneous fat*. The diminutive *cuticle* has various meanings to do with skin, including *epidermis*. Another Latin word for skin or hide was **pellis**, from which we get *pellagra*, the deficiency condition caused by lack of *niacin*, i.e. vitamin B3, the precursor for the coenzyme NAD. One of the symptoms of pellagra is dermatitis. **Agra** (Gk.) 'disease' also shows up in **podagra** 'gout in the feet' and **chiragra** 'gout in the hands'.

DIMINUTIVES are nouns derived from other nouns and altered by the addition of a suffix to indicate diminished size or status, from **deminuere** (L.) 'to lessen' from **minuere** 'make small'. Diminutives seem to occur in most languages. In the case of Latin, diminutives consist of parent words typically modified by the suffix **-ulus**. An example is the name of the emperor Gaius, who ruled the empire during 27–31 CE. As a small child, he accompanied his father Germanicus on military campaigns, during which he would be dressed up in a miniature soldier's uniform, including miniature army boots (**caligae**). He thus acquired the nickname Caligula, which stuck and by which he is remembered. Its English equivalent would be Bootikins.

As with English in general, many scientific and technical diminutives have come across from French. Some of these are recognizable by the suffix *-et* or *-ette*, suffixes in French which distinguish male and female nouns. Examples of the latter are *pipette*, the laboratory tool for dispensing known volumes of liquids, derived from **la pipe**, and *statuette* from **la statue**. Diminutives derived from French words and terminating in *-et* are not so obvious: *pocket* derived from the diminutive of the OF word **poke** or **poque**, meaning 'bag'. The poke here is the origin of the saying 'sold a pig in a poke', currently meaning to settle a deal without going into the details or to be the victim of a con man. The original idea was that when you opened the bag, you discovered you had been 'sold a pup'; *bullet*: from the diminutive of French **boule** meaning 'ball', describing an item with a similar function but much smaller than a cannonball.

Another diminutive suffix, characteristic of Germanic languages is **-ling**, as in *sapling, sibling, gosling, duckling.*

Yet another such suffix is *-let*: *bracelet* from **bras** (F) 'arm', *gauntlet* from **gant** (F) 'glove', and some home-grown examples: *leaflet, booklet, piglet,* and first defined in German, *platelet*—see below.

Alveoli	the tiny air sacs in the lungs, diminutive of **alveus** 'cavity'.
Bacillus	a type of rod-shaped bacterium, in Latin, **baculum** 'staff ' or 'stick', **bacillus** 'small staff'.
Brittunculi	'little Brits'—a contemptuous term used by the Romans to describe the natives of Britannia.[36]
Carbuncle	from **carbunculus** 'small coal', diminutive of **carbo** 'coal'. The Romans referred to various red-coloured precious stones such as garnets as **carbunculi**, imagining them to resemble hot coals in a fire. Medically, *carbuncle* refers to inflammation of the skin resulting from infection by *Staphylococcus aureus*. The term *carbuncle* is obsolescent; the condition would now be described as a boil.
Cerebellum	'small brain', from **cerebrum** (L.) 'brain'. The cerebellum lies at the base of the brain and has the appearance of a separate structure, thus accounting for the 'small brain' label. Among its functions is the coordination of motor activity.
Coagulate	from **coagulare** 'to cause to coagulate or curdle', diminutive of **coagere** 'to cause to run together', from **co-** + **agere** 'to impel'.
Cuticle	skin or epidermis, from **cuticula**, diminutive of **cutis** 'skin'.
Fibril	small fibre, from Latin **fibrilla**, diminutive of **fibra** 'fibre'.

Flagellum	the whiplike structure that confers motility on single-cell organisms such as bacteria, protozoa, spermatozoa, diminutive of **flagrum** 'whip'.
Formula	diminutive of **forma** 'shape, form, kind'.
Globulin	from **globulus** 'globule', diminutive of **globus** 'round object'.
Granulocyte	a type of white blood cell containing granular inclusions in its cytoplasm, from **granulum** 'small seed', diminutive of **granum** 'seed'.
Homunculus	diminutive of **homo** 'man', a supposed microscopic but fully formed human from which, in former times, it was believed a foetus would develop.
Jugular	referring to the neck or throat, diminutive of **jugum**, the wooden crosspiece or yoke placed across the neck of a pair of bullocks and attached to a plough or wagon, also horse collar.
Lamella	a thin layer of bone or tissue, diminutive of **lamina** 'a layer'.
Meniscus	the curved upper surface of a liquid in a tube, which reminded the ancients of the shape of a new moon; from **meniskos** (Gk.) 'crescent', diminutive of **mene**, 'moon'.
Molecule	from modern Latin **molecula**, derived from **moles** 'mass'.
Muscle	from **musculus** 'small mouse', diminutive of **mus** 'mouse'.
Nucleolus	a differentiated region of the nucleus which is the site of ribosomal RNA synthesis; diminutive of **nucleus** 'inner part, kernel'.
Organelle	such as the mitochondrion, the diminutive of **organum**, from the Greek **organon**: an implement for making or doing things
Particle	from **particula** 'little part', diminutive of **pars, part-**, hence *particulate*: composed of particles; cf. *article* and *articulate*.
Pipette	from **pipe** (F) as noted above. In most labs, *la pipette* has undergone a sex change and become *le pipet*.

Platelet	small disc-shaped anucleate cell fragment, critical for blood clotting—see also under **BLOOD.**
Reticulum	diminutive of **rete** (L.) 'net', as in *endoplasmic reticulum*: the membranous network in the cytoplasm of eukaryotic cells. The *ruminant stomach* is described as consisting of the *rumen, reticulum, omasum*, and *abomasum*.
Testicle	organ producing spermatozoa, from **testis** (L.); plural: **testes,** which is the answer to the question 'What's that hanging underneath him?' sometimes asked by city children on farm visits when they first encounter a ram.
Tubercle	a protuberance; in a medical context, a small nodular lesion on the lung, characteristic of tuberculosis caused by the tubercle bacillus, from **tuber** (L.) 'swelling or lump'.
Tubule	as in tubules of the kidney, from **tubulus**, diminutive of **tubus** 'tube or pipe'.
Vascular	from **vasculum,** diminutive of **vas** 'vessel'; in animals, vascular tissue is that enriched in blood vessels. The *vascular system* includes the heart, arteries, veins, and capillaries. Other vascular systems are the lymphatic system as well as that in *vascular plants* that conducts water and nutrients from the roots.
Ventricle	cavity of an organ, e.g. the left and right ventricles of the heart; also cavities in the brain filled with cerebrospinal fluid; from **ventriculus**, diminutive of **venter** 'belly'.
Vesicle	fluid-filled cavity or sac from **vesicula**, from **vesica** 'bladder'.

DROMOS (Gk.): 'race, running'. PALINDROME: **palin** (Gk.) 'back, backwards' + **dromos**. Strictly speaking, a palindrome consists of a word or series of words or letters which reads the same in both directions such as the line attributed to Napoleon: 'Able was I ere I saw Elba.' The rule with palindromes is that their sense declines exponentially with their length. It turns out that molecular biologists are also interested in palindromes, this time in DNA sequences. A DNA palindrome is where the two DNA strands, each read in the 5' to 3' direction have the same sequence, for example:

5' GGCAGTACTGCC 3'
3' CCGTCATGACGG 5'

These of course are not true palindromes but are quite reasonably described as *palindromic*. They are more than curiosities because in some cases, they are telling us about sequence rearrangements that have occurred, i.e. the history of a particular sequence; in other cases, they represent binding sites for regulatory proteins, which often occur as *dimers* which bind to the same sequence on each strand. See the entries for the **Y CHROMOSOME**, and for **CRISPR** and **GENE EDITING**, where SPR stands for short palindromic repeats. There is also an obscure term, *enantiodromic*; you will recall that *enantiomers* are molecules that are mirror images, the prefix **enantio**- meaning 'opposite'. So you *could* say that the two strands of DNA are *enantiodromic* because they run in opposite directions. Well, you could say that, but not many people would know what you were talking about if you did.

PRODROME: relating to the period between the appearance of first symptoms and the full development of disease, used in connection with neurodegenerative and other diseases that are slow to develop. In a somewhat different context, 'this prodrome period' was used by a political commentator to refer to the period between the election of President Trump and his swearing in.

SYNDROME: **syn** 'together' + **dromos**: the concurrence or running together of symptoms. In medicine, it refers to the association of a number of clinically recognizable features that together are characteristic of a particular condition. Fish people have also found a use for the **dromos** root: *anadromous*, fish that spend most of their life in the sea and migrate to fresh water to breed, i.e. they swim upstream (**ana**- 'up'), e.g. many species of salmon; *catadromous*: fish that do it the opposite way round and migrate to the sea to breed. Other words with the **dromos** root: *hippodrome, velodrome, aerodrome, dromedary*. The Latin to run is **currere**, *pp.* **cursus**, from which we get *current, cursory, discourse, excursion, incur*, and *precursor*.

DYSLEXIA: **dys** + **lexis** (Gk.) 'speech or word'. Dyslexia is defined as a disorder causing difficulty in learning to read despite conventional instruction and normal intelligence. According to the *New Oxford American*

Dictionary, dyslexia was coined in German from **dys** 'difficult' + Greek **lexis** 'speech or word' (apparently by confusion of Greek **legein** 'to speak' and Latin **legere** 'to read'). While the Greek verb does mean 'to speak', according to Liddell and Scott, the authorities on ancient Greek, it can also mean 'to read' when present in a compound, so this may well have been the rationale when Rudolf Berlin came up with *dyslexia* in 1887. *Lexicon*, meaning 'language or vocabulary', is derived from the same root, usually describing the words typical of a particular discipline or region. A dictionary maker is a *lexicographer*, famously defined by Dr Samuel Johnson in his English dictionary in 1755 as 'a harmless drudge'. There is also *alexia* (L.), meaning 'inability to see words or to read'. On the other hand, **legere** (*ppp.* **lectum**) has multiple meanings depending on context, mainly 'to gather, collect, select', but it can also mean 'to read aloud or recite' and is the root for *lectern*, which originally referred to a desk in a church from which lessons were read. **Legere** is also the root for *dialect*, which brings to mind the line 'A language is a dialect with an army.'[37]

ELECTRO- elektron (Gk.) 'amber'. This prefix, used to indicate an involvement of *electricity* or of *electrical charge* derives from the fact, known from ancient times, that rubbing amber with a cloth generates static electricity. In Latin, the word for amber, adopted from the Greek, was **electrum**.

ELECTROLYTE: electro- + **lutos** 'released'. Electrolytes are chemical compounds which when dissolved in water dissociate into *positive* and *negative ions*, e.g. $NaCl \rightarrow Na^+ + Cl^-$. In a physiological context, the important electrolytes are sodium, potassium, calcium, magnesium, chloride, phosphate, and bicarbonate. Various mechanisms involving *antidiuretic hormone, aldosterone*, and *parathyroid* hormone regulate overall electrolyte *homeostasis*, which is critical for maintenance of blood pH, nerve and muscle function.

Cells of higher organisms maintain high levels of intracellular potassium and low levels of sodium compared to the extracellular medium, usually the blood. This is due to activity of the *sodium-potassium pump*, which consists of a transmembrane enzyme known as the Na^+/K^+ ATPase, which for each cycle exports 3 Na^+ while importing 2 K^+. See under **SODIUM** and **POTASSIUM** heading.

Cholera, characterized by vomiting and diarrhoea, and otherwise a life-threatening disease, can be simply and effectively treated by replacing water and electrolytes. The name derives from **khol**e 'bile' because in classical times, an excess of bile was believed to be the cause. See **BILE** heading.

ELECTROPHORESIS: **electro-** + **phoresis** 'being carried'. Compounds that carry a net charge migrate in an electric field. Electrophoresis is the term used to describe the application of this effect to the separation of charged compounds such as proteins or polynucleotides, as well as amino acids, peptides, oligonucleotides, etc. By using a supporting medium, such as a *polyacrylamide* gel at an appropriate concentration that has a *sieving effect*, electrophoretic separations can be achieved on the basis of size as well as charge. ELECTRODE: **electro-** + **hodos** (Gk.) 'way'. An electrode is an electrical conductor, usually metal, used to make contact with a liquid part of a circuit. Michael Faraday, in 1834, contributed an article, 'On electrical decomposition' to the *Philosophical Transactions of the Royal Society* in which he introduced the terms *electrode*, *anode*, *cathode*, *anion*, *cation*, *electrolyte*, and *electrolyse*. The **hodos** root also shows up in words such as *method*, *diode*, *episode*, *exodus*, *odometer*, and *period*.

ELEMENT: the English word comes directly from the Latin **elementum,** which as usual had various meanings depending on context but included the idea of something rudimentary, a first principle from which other things developed. The concept of elements goes back to Aristotle and, beyond: everything was supposedly explained by the various combinations of four elements: fire, air, earth, and water. This was the theoretical basis for alchemy. In protest at the ill-defined use of the word by alchemists, Robert Boyle in *The Sceptical Chymist* (1661) defined it as follows: 'And to prevent mistakes, I must advertise you that I now mean by elements . . . certain primitive and simple or perfectly unmingled bodies.' [38] Lavoisier in 1787 made a provisional list of thirty substances that appeared to be unmingled, a forerunner of the periodic table. The Greek word for element was **stoikheion**, which has come across into English along with **metron** 'measure' as *stoichiometric*, meaning the whole number ratios in which atoms and molecules react.

ELUTE: (L.) **eluere, elutum** 'to wash out or wash off'. A purification procedure may involve binding to an ion-exchange column a mixture of compounds which are then *eluted* by, for example, washing the column with solutions of increasing salt concentration. Purification is achieved when the components of the original mixture elute at different salt concentrations. Here the eluting solution is the *eluent*, and the solution emerging from the bottom of the column is the *eluate*. The related verb **diluere** had various meanings in Latin such as 'to dissolve', 'to wash away', and 'to dilute.' The stem of these verbs corresponds to the Indo-European **leue-**, meaning 'wash', which is also the stem for *lye*—an obsolete term for sodium and potassium hydroxides, but still used in industry.

ENDOPLASMIC RETICULUM (ER) is an *organelle* in the cytoplasm of eukaryotic cells, consisting of a continuous network of *membranous tubules* also referred to as *sacs* or *cisternae*, the latter meaning a fluid-filled cavity. The ER was first observed in detail in 1945 upon development of the electron microscope. It was termed *endoplasmic* because it tends to be absent from the margins of the cell, instead occupying the central or endoplasmic portions of the cytoplasm. *Reticulum* means network. All cells have ER except for red blood cells which, at least in mammals, also lack nuclei—seemingly to provide as much space as possible for haemoglobin.

In electron micrographs of cells making large amounts of *secreted proteins*, such as liver cells or those from lactating mammary gland, many ribosomes which are in the process of synthesizing proteins are observed attached to the ER. Such ER is described as *rough ER* as distinct from *smooth ER*, which is involved in lipid synthesis and metabolism together with other reactions catalysed by lipid-bound enzymes. The relative proportions of the two types of ER vary depending on the type of cell. The *sarcoplasmic* reticulum of muscle is smooth ER. See also entry for **MUSCLE**.

ENTERO- **enteron** (Gk.) 'intestine'.
ENTEROCYTES: **entero-** + **kutos** 'vessel'. Major cell type of the intestinal epithelium, adapted for absorption, that also generate *chylomicrons* and regulate the level of iron in circulation. See under **IRON**.

ENTEROPEPTIDASE: a proteolytic enzyme secreted by cells that line the duodenum and which, by cleaving a single peptide bond, activates trypsinogen as the zymogen enters the duodenum from the pancreas. Trypsin then activates other pancreatic zymogens.

ENTERIC BACTERIA can refer either to normal gut flora or to pathogenic bacteria.

DYSENTERY: a term implying a somewhat more serious infection.

GASTROENTERITIS: **gaster** 'stomach' + **enteron**, infection affecting both the stomach and intestines.

COELENTERATE: **koilos** 'hollow' + **enteron**, a group of marine invertebrates that includes jellyfish, corals, and sea anemones. These were formerly grouped together in the phylum Coelenterata but are now divided into Ctenophora and Cnidaria.

ENTEROHEPATIC circulation of bile acids. See under **CHOLESTEROL**.

PARENTERAL (*adj.*): **par-** (Gk.) 'beside', in the sense of 'as an alternative' + **enteron**: involving the introduction of a substance into the body other than by the alimentary tract, e.g. by injection or infusion.

ENZYMES: **en-** 'in' + **zyme** 'yeast'. Nearly all enzymes are proteins, with the exception of a few RNA-based enzymes, *ribozymes* (see below) that are regarded as survivors from a much earlier RNA world. Enzymes are *catalysts*, and this means they accelerate chemical reactions but do not alter the equilibrium between reactants and products. Instead, they lower the *activation energy* for the reaction. The activation energy can be understood as the energy required to convert the *reactants* or *substrates* into what is called the *transition state*, which corresponds in structure to neither substrate nor *product*; it is an activated intermediate along the reaction pathway between substrate and product. Enzymes facilitate reactions through multiple binding interactions that bring substrates together in a configuration that is favourable to formation of the transition state and determines enzyme specificity. The business end of an enzyme, where substrates bind, is called the *active site*. The active site of an enzyme consists of amino acid side chains arranged in a specific 3D arrangement. It is the energy released by the multiple binding interactions between enzyme and substrates and, ultimately, by the transition state that overcomes the activation energy barrier of the reaction. Enzymes are amazingly efficient.

For some enzymes, the factor limiting the rate at which they catalyse reactions is the rate at which substrate and enzyme encounter each other, i.e. the rate is *diffusion controlled*. Such enzymes are said to be *catalytically perfect* in that every encounter between substrate and enzyme is productive.

RIBOZYMES: It had long been assumed that all enzymes had to be proteins, a belief reinforced as X-ray structures of enzymes were obtained with bound substrates and inhibitors, thus enabling a detailed understanding of enzyme mechanisms. It came as a surprise therefore when it was discovered that RNA itself can have catalytic activity. The initial finding was that an RNA sequence, termed a type I intron, located within a ribosomal RNA transcript in the tiny protozoan *Hymenoptera*, was capable of autoexcision. Self-splicing type I introns have now been found in bacteria, lower eukaryotes, and higher plants.

This was followed by the discovery that *RNaseP*, which consists of a *ribonucleoprotein* and functions in the cleavage of transfer RNA precursors, was also able to catalyse this reaction when the protein component had been removed. It has since been further shown that the critical reaction of protein synthesis catalysed on the ribosome by *peptidyl transferase*, the addition of an incoming amino acid to the end of a growing peptide chain, is also RNA-catalysed by ribosomal RNA nucleotides. Thus, three of the key functions involved in modern protein synthesis can be performed by RNA. This leads to the irresistible idea that our current DNA-based world was preceded by an RNA world in which RNA performed the genetic function of DNA as well as the catalytic function of enzymes.[39]

DISCOVERY of ENZYMES. The first investigation of what we would now call enzyme activity was in a paper by Payen and Peroz in 1833 reporting that the malting of barley involved the release of sugar from the grains of the barley as it began to germinate during the malting process. They found that an aqueous extract of malt, when precipitated with alcohol and redissolved, was able to convert starch into sugar. This activity they called *diastase*, from **diastasis** (Gk.) 'separation', on account of its ability during malting to separate sugar from the husk of the barley seed. About the same time, in 1834, it was observed that gastric juice could digest food in a test tube, and Schwann (1810–1882) in Berlin showed in 1836 that this activity could be precipitated and called it *pepsin* after the Greek **pepsis** 'digestion'. For a time, all such activities were referred to as diastases or alternatively as *ferments* and sometimes as 'influences'.

The word *catalysis* was introduced in 1836 by Berzelius (1779–1848) to explain these early observations of enzyme action. One meaning of the Greek **kataluein** is 'to dissolve', so possibly Berzelius was influenced by the ability of the malt extract to produce soluble sugar from insoluble starch and the fact that the only examples of enzyme action then known were degradative, i.e. *catabolic.* Berzelius was also aware of the ability of platinum to facilitate the conversion of methane and alcohol into other compounds, as indicated by the quote below. It is thought however that it was these observations on enzymes that led him to the concept of a catalyst.

From the *OED*: 1836, Berzelius in *Edin. New Phil. Jrnl.* XXI, 223: 'Many bodies ...have the property of exerting on other bodies an action which is very different from chemical affinity. By means of this action they produce decomposition in bodies, and form new compounds into the composition of which they do not enter. This new power, hitherto unknown, is common both in organic and inorganic nature ...I shall ...call it catalytic power. I shall also call Catalysis the decomposition of bodies by this force.'

In another article in 1837, Berzelius speculated with amazing foresight: 'When we turn with this idea to living Nature, an entirely new light dawns for us: It gives us good cause to suppose that in living plants and animals thousands of catalytic processes are taking place between the tissues and the fluids, producing the multitude of dissimilar chemical compounds for whose formation from the common raw material, sap or blood, we had not been able to think of any cause, but which in the future we shall probably find in the catalytic power of the organic tissue of which the organs of the living body consist.[40]

In 1860, the French chemist Berthelot ground up yeast and obtained an activity, termed *invertase*, able to hydrolyse sucrose. Kuhne in 1878 introduced the term *enzyme* from **en** + **zyme** (Gk.), meaning literally 'in yeast', the idea originally being to distinguish the yeast activity from others such as diastases. In 1898, Duclaux suggested that the -*ase* suffix of diastase be used for all enzymes. Buchner (1860–1917) showed in 1897 that an extract of yeast cells was able to carry out fermentation, thus demolishing the idea promoted by Pasteur that fermentation could only occur within the intact living cell. Buchner's experiment marks the beginning of the subject that became known as biochemistry and of its divergence from organic chemistry. Biochemists ever since have been breaking cells open, purifying components and recombining them to understand *in vivo* processes such as how DNA is replicated, proteins are synthesized, etc.

COFACTORS, COENZYMES, and PROSTHETIC GROUPS. All of these terms refer to usually low-molecular-weight molecules (often derivatives of *vitamins*) that are necessary for the function of an enzyme or other type of protein. Apart from low-molecular-weight organics, the other group of cofactors, also known as *micronutrients*, are *metal ions* such as Mg^{++}, Mn^{++}, Fe^{++} and several other less-common cations such as Co^{++} in vitamin B12 and Mo^{++} in the *nitrogenase* complex. The term *prosthetic group*, which has the same derivation as *prosthesis*, in Greek meaning 'addition', in English meaning an artificial body part, is used to indicate that the small molecule is tightly bound to the protein. Even then, there are two possibilities: is it covalently or non-covalently bound? Examples of enzymes with prosthetic groups non-covalently but tightly bound include numerous *flavoenzymes* containing *flavin adenine dinucleotide* (FAD) or *flavin mononucleotide* (FMN), including *pyruvate dehydrogenase* and the citric acid cycle enzyme *succinate dehydrogenase*. *Biotin*, which is required for carboxylation reactions, forms an amide bond with a lysine residue of carboxylase enzymes. The haem group of haemoglobin is an example of a prosthetic group tightly bound non-covalently.

FAD and FMN together with *nicotinamide adenine dinucleotide* (NAD⁺) and its phosphorylated derivative NADP⁺ are able to accept and donate electrons, so they can be described as *redox cofactors*. Coenzymes such as coenzyme A and NAD that bind to enzymes where they serve as either donors or acceptors of functional groups and then dissociate are really *co-substrates*. Proteins lacking their prosthetic groups are referred to as *apoproteins* or *apoenzymes* (**apo-** 'from' or 'away'). See also the entry for **VITAMINS,** most of which are precursors of coenzymes. Structures for NAD⁺, FAD and FMN can be found at https://en.wikipedia.org/wiki/nicotinamide_adenine_dinucleotide. Likewise for flavin_adenine_dinucleotide and flavin_mononucleotide.

ENZYME INHIBITION: **inhibere** (L.) 'to hinder'. Enzyme *inhibitors* bind to enzymes and decrease or abolish their activity. A common, naturally occurring mode of enzyme inhibition is known as *allosteric inhibition*, **allos** (Gk.) 'other' + **steros** 'solid', but here meaning 'position'. This is typically observed in a biosynthetic pathway, such as that leading via a number of intermediates to an amino acid or a nucleotide. Here the end product is found to inhibit the enzyme that catalyses the first step in the pathway by binding to a site on the enzyme (termed an *allosteric site*) that is distinct from its *active site*. This prevents wasteful synthesis of intermediates when the concentration of the final product is adequate.

It is hard to exaggerate the significance of enzyme inhibition for human health. The correlation between high blood cholesterol levels and mortality from heart disease led to attempts to lower the rate at which cholesterol is synthesized *in vivo*. The *committed step* in cholesterol synthesis is *HMG CoA reductase*, and inhibitors of this enzyme, the *statins*, are widely prescribed by physicians. See entry for CHOLESTEROL SYNTHESIS. The life cycle of the *retrovirus* HIV involves the synthesis of various viral proteins and enzymes as a *multidomain* protein or *polyprotein*. This is subsequently acted on by an HIV-encoded protease that releases the individual proteins. This protease is critical for HIV replication and inhibitors of its action are effective antivirals. Gleevec is an inhibitor of the Abl *tyrosine kinase* and has transformed the survival outlook for most sufferers of *chronic myeloid leukaemia* (CML), the cancer caused by a *chromosomal translocation* that results in dysregulation of Abl kinase expression. See entry for CHRONIC MYELOID LEUKAEMIA under **CANCER**. Enzymes are also the targets for many *antibiotics*. A criterion to be met is that the enzyme is present in the target bacterium but not in humans, or if it is, then it is sufficiently different so as to be unaffected by the antibiotic.

ETHYL- is the name of the CH_3CH_2 *radical* as in ethyl alcohol CH_3CH_2OH, ethylamine $CH_3CH_2NH_2$, the ester ethyl b acetate $CH_3COOCH_2CH_3$, and many other compounds. The term *radical*, from the Latin **radicula**, the diminutive of **radix, radicis** meaning 'root', was introduced to describe a group of atoms that remained intact through a variety of reactions. In the early days of chemistry, the term *ether*, from **aither** (Gk.) 'air' or 'sky', was applied to any volatile compound. In 1834, Liebig suggested *ethyl* as a name for the CH_3CH_2 radical, this being a combination of **eth-** from ether and **yl** from the Greek ὑλη, meaning 'matter'. In 1836, Berzelius followed suit by naming methyl, from the Greek **methy** meaning 'wine', as the radical in wood spirit, i.e. methyl alcohol or methanol.[41] ETHYL ALCOHOL or ETHANOL is the chemical generally known as alcohol, probably the first recreational drug. It was known in mediaeval times as *spirit of wine* and by the alchemists as **aqua ardens,** the burning liquid. Paracelsus (1493–1541) called it alcohol, based on an Arabic loan word, *al-kohl,* and this was the name eventually adopted. Alcohol has many uses apart from those associated with hangovers and is readily available as *denatured* alcohol that contains compounds added to make it undrinkable, such as pyridine or methanol as in *methylated spirits.* This avoids the taxes levied on alcohol for consumption. Alcohol is also a useful laboratory reagent. The traditional first step in DNA isolation is adding alcohol to a tissue homogenate and then winding out the DNA on a glass rod. The Cohn method for the large-scale fractionation of blood plasma proteins, developed during World War II, involved varied alcohol concentrations, pH, and ionic strength at low temperature. A 70/30 v/v mixture of alcohol and water is a potent *antiseptic.*

The general purpose ethanol reagent used in laboratories is not 100 per cent pure but 95.6 per cent, usually referred to as '95 per cent'. This comes about because the *fractional distillation* of an alcohol-water mixture yields an *azeotrope* (**a-** 'no', **zein,** 'to boil', **tropein** 'to turn', meaning 'no effect of boiling') which occurs when the composition of the vapour is the same as that of the liquid mixture. Mixtures of compounds that yield azeotropes are not uncommon. In situations where pure alcohol is required, the remaining water can be removed using minerals known as *zeolites,* which adsorb water and act as *molecular sieves.*[42] Powdered *cellulose* can also be used. Pure, 100 per cent ethanol is usually described as *absolute alcohol.* Most ethanol produced today is used as motor fuel, either directly or as an additive, and the water may or may not be removed depending on the formulation.

FATTY ACIDS when found on their own are referred to as *free fatty acids.* They consist of a carboxyl group attached to a carbon chain of variable length, which may be *saturated* or *unsaturated.* The general formula for saturated fatty acids is $CH_3\text{-}(CH_2)_n\text{-}COOH$. For naturally occurring fatty acids, n is usually an even number, reflecting the mode of synthesis in which two-carbon units are added sequentially. Fatty acids being insoluble in water are transported in the blood bound to *serum albumin*. Most naturally occurring fatty acids contain between four and twenty-four carbon atoms.

BUTYRIC ACID: **butyrum** (L.) 'butter', related to the Greek **bouturon** 'cow cheese', from *bou* 'cow' and **turon** 'cheese', the root for the amino acid tyrosine, which was first isolated from cheese. Formula: $CH_3(CH_2)_2COOH$. Systematic name: *butanoic acid*.

CAPROIC ACID: *hexanoic acid*, $CH_3(CH_2)_4COOH$, i.e. a 6-carbon carboxylic acid, a colourless, oily liquid with an odour like that of goats.

CAPRYLIC ACID: *octanoic acid*, $CH_3(CH_2)_6COOH$, i.e. an 8-carbon carboxylic acid, an oily liquid with a somewhat unpleasant smell and taste.

CAPRIC ACID: *decanoic acid*, $CH_3(CH_2)_8COOH$, i.e. a 10-carbon carboxylic acid. These three acids are examples where systematic names make it much easier to remember which is which. The *glyceryl esters* of the above three acids account for 15 per cent of goat milk fat. CAPRINE: **caprinus**: 'relating to goats' (cf. ovine, bovine). The domesticated goat is *Capra hircus*. *Hircus* in Latin had various meanings, from he-goat or buck to a goatish smell such as the rank smell of the armpits, to an epithet applied to a filthy person.

LAURIC ACID: C12 saturated fatty acid. Formula: $CH_3(CH_2)_{10}COOH$. Systematic name: *dodecanoic acid*. It only occurs in significant amounts in some vegetable oils including laurel oil, obtained from berries of the laurel tree *Laurus nobilis*. An ionic detergent commonly used in the laboratory to disperse membranes and protein aggregates is *sodium lauryl sulphate*, more commonly referred to as *sodium dodecoyl sulphate* (SDS). The term *detergent* comes from the Latin **tergere**, also **detergere**, both meaning, depending on context, 'to wipe clean' or 'to cleanse'. Leaves from the laurel tree, also known as the bay tree, are used in cooking, while laurel wreaths were worn as a sign of status in ancient Greece and by victorious generals in Rome. This association with status lingers in the related words laureate, as in *poet laureate* and *baccalaureate*.

MYRISTIC ACID: C14 saturated fatty acid. Formula: $CH_3(CH_2)_{12}COOH$. Systematic name: *tetradecanoic acid*. It is named from the nutmeg *Myristica*

141

fragrans, which in turn takes its name from **myristikos** (Gk.) meaning 'fragrant'. Nutmeg butter is 75 per cent trimyristin, the triglyceride of myristic acid. MYRISTOYLATION refers to the process whereby a myristoyl group is covalently attached to a protein via an amide linkage to the N-terminal amino acid, usually a conserved glycine residue, after removal of the initiating methionine. Proteins modified in this way are often involved in signal transduction pathways, and addition of the *myristoyl* group facilitates their interaction with membranes or with other proteins. Pathogens target this signalling landscape. A *Shigella* strain *demyristoylates*, via a cysteine protease, an array of proteins, including those involved in cell growth, *signal transduction, autophagosome* maturation, and organelle function. The protease removes the myristoyl group and the conserved glycine, another chapter in the arms race between bacteria and their hosts.

PALMITIC ACID: C16 saturated fatty acid. Formula: $CH_3(CH_2)_{14}COOH$. Systematic name: *hexadecanoic acid*. The most common fatty acid found in living organisms, it is the major fatty acid component of palm oil, hence the name. In World War II, aluminium salts of palmitic and naphthenic acids provided an early version of napalm, and the name as well.

STEARIC ACID: **stear, steat-** (Gk.) 'fat, tallow'. C18 saturated fatty acid. Formula: $CH_3(CH_2)_{16}COOH$. Systematic name: *octadecanoic acid*. Occurs mainly as a constituent of triglycerides in animal fats and vegetable oils. The same Greek root occurs in *steatohepatitis*, also known as fatty liver.

OLEIC ACID: monounsaturated C18 fatty acid, systematic name: 9-octadecaenoic acid (counting from the carboxyl group). This indicates that the double bond occurs between the ninth and tenth carbon atom. Formula: $CH_3(CH_2)_7CH=CH(CH_2)_7COOH$. Oleic from **oleum** (L.) 'oil'.

LINOLEIC ACID: $C_{17}H_{31}COOH$ present in triglycerides in linseed oil from the flax plant and in other oils and essential in the human diet. Systematic name: 9, 12-octadecadienoic acid. The flax plant is *Linum usitatissimum*. **Linum** was the Latin word for 'flax' and has given rise to the English words *linen* and *line*, which originally meant 'rope'. The species name means 'often used' or 'very useful', which reflects its use as a source of both food and fibre.

LINOLENIC ACID: $C_{17}H_{29}COOH$, 9,12,15-octadecatrienoic acid. Linseed oil can contain up to 70 per cent of these two polyunsaturated fatty acids. Linoleic and linolenic are the two *essential fatty acids*. 'Essential' means 'must be supplied in the diet', and this is because mammals lack the enzymes required to introduce double bonds beyond C-9 in fatty acids.

Fatty acids important for human health are often referred to as ω-3 and ω-6 fatty acids. Carbon atoms in fatty acids are normally numbered from the carboxyl end; however, it is towards the other end, the ω-end or methyl end, that is subject to variable *desaturation*. The term ω-3 indicates that the first double bond involves the third and fourth carbons from the ω end. Thus, linoleic acid is an ω-6 fatty acid and linolenic acid is an ω-3. These two C18 fatty acids are the starting points for the synthesis of other, longer, and more unsaturated fatty acids.

The C20 polyunsaturated fatty acids and their derivatives are referred to as *eicosanoids*, Greek: **eikosi** 'twenty'. Linolenic gives rise to *eicosapentenoic acid* with five double bonds, while among various C20 ω-6 fatty acids, the important one is *arachidonic acid* 5,8,11,14-eicosatetraenoic acid synthesized in humans from linoleic acid. It is named for its occurrence in peanut oil, also known as groundnut oil since peanuts, the product of *Arachis hypogaea*, develop underground as indicated by the species name. Arachidonic is the precursor for a number of C20 hormones which include *leukotrienes*, *prostacyclin*, *prostaglandins*, and *thromboxanes*, the result of a variety of modifications including ring formation. These, also known as *prostanoids*, have short half-lives, act locally (i.e. in an *autocrine* or *paracrine* manner), and exert different effects in different tissues by binding to specific receptors. *Aspirin* inhibits one of the initial steps in the modification of arachidonic acid, which accounts for the multiple pharmacological effects of aspirin.

The C22 fatty acid 4,7,10,13,16,19 docosahexaenoic acid (DHA) is an ω-3 fatty acid, synthesized from linolenic acid and is stated to be the primary structural component of the human brain. The **dokosa-** of DHA comes from **dokos** (Gk.), a load-bearing beam in the roof or floor of a house.

Two further points relating to the nomenclature of fatty acids: because at physiological pH, the carboxyl groups will be in the salt form, i.e. ionized to COO^-, strictly speaking therefore, in their *in situ* context they should be referred to as palmitate, linoleate, linolenate, arachidonate, etc. Also, for unsaturated fatty acids, there are for each double bond two possible stereoisomers, *cis* and *trans*. The naturally occurring isomers are all *cis*.

FLOW: the Greek **rhein** 'to flow' has provided the stem for several words in English. The most obvious is *rheostat*, from **rhein** + **stasis** 'standing' or 'stoppage', the electrical instrument which controls the flow of current by varying the resistance, cf. *haemostat, thermostat*. Similarly, the subject of *rheology* has to do with the flow of matter, particularly the flow of liquids. Other forms of the verb are distinguished by addition of an extra consonant as in the medical terms below. The corresponding Latin verb is **fluere** 'to flow'—see **FLUX** heading below.

Haemorrhage or 'a flow of blood', from **haimorrhagia**; *haemorrhoids* from **haimorrhois**; *gonorrhoea* from **gonorrhoia** where infection may be accompanied by discharge (**gone**: 'generation' or 'seed' cf. *gonad*); *diarrhoea* from **diarrhoia** (**dia-** 'through', hence 'to flow through'). A semi-obsolete medical term is *catarrh*, derived from **katarrhous** (**kata-** 'down', hence 'flowing down'), a profuse discharge from the nose accompanying a cold, formerly supposed to 'run down' from the brain. Let's settle on *rhinorrhoea* **rhinos** (Gk.) 'nose' for a runny nose. *Steatorrhea* refers to excessive excretion of fat in the faeces due to a lack of bile acids, named for *stearic acid*.

The noun corresponding to **rhein** is **rheos**, translated as 'stream', while a variant of **rheos** was **rheuma** with the same meaning which has given rise to modern *rheumatism*, originally thought to be due to the internal flow of watery humours. *Rheum* is defined by the *OED* as watery matter secreted by the mucous glands or membranes. An *OED* quote dated 1897 from Allbutt's *A System of Medicine*: 'The rheumatic pain is attributed to rheum flowing down from the brain and settling in the affected area.'

FLUX, FLUORINE, and FLUORESCENCE: these terms derive ultimately from **fluere, fluxi, fluxum** (L.) 'to flow'. The mineral fluorspar, or *fluorite*, CaF_2, was named for its use as a *flux* in the smelting of ores, where its function is to lower the melting temperature and decrease the viscosity of slags. This use for fluorite is first mentioned in 1529. The name of the element *fluorine* was suggested by Humphry Davy in 1810 for the then-unknown element present in fluorite, on the basis of experiments that suggested similarities with chlorine. Fluorine was not isolated until 1886. Similarly, the term *fluorescence* was devised to describe a phenomenon characteristic of some fluorites containing impurities. Fluorescence is exhibited by compounds that, when illuminated with light of a certain wavelength, emit light of a longer wavelength. This occurs when the incident light causes an orbital electron to be excited to a higher energy level and to then return to the ground state with the emission of light. Many marine organisms, when viewed under UV light, fluoresce in the visible spectrum with a variety of colours. A protein isolated from the jellyfish *Aequoria victoria*, termed the *green fluorescent protein* (GFP), has been adapted as a protein tag for many different research applications such as localizing particular proteins within cells. The GFP *chromophore* is a polycyclic compound, *coelenterazine*, which is excited by light at around 395 nm and emits light of around 509 nm. A variety of such coloured tags are now available.

FLUOROACETATE: A poison, marketed as 1080, which inhibits the citric acid cycle by forming *fluorocitrate*. Some poisonous plants, particularly *Gastrolobium* species in Western Australia, synthesize fluoroacetate, evidently as a defence against herbivores. It is claimed that some native species such as possums can consume these plants without problems, whereas introduced species such as cattle are susceptible. A similar situation is reported from the Mojave Desert, where native wood rats can graze on the poisonous *creosote bush* partly because their livers can detoxify the poisons and because their gut microbes confer resistance to the toxins.[43] Similar detoxification mechanisms may operate for fluoroacetate. Other words from the **fluere** root: *effluent, confluent, fluent, fluid, reflux, superfluous.*

INFLUENCE and INFLUENZA. The word *influence* harks back to the days of *astrology* when it was believed that an ethereal *fluid* streamed down from the stars and determined human destiny. Later, it was viewed as an occult force as in Shakespeare's *Hamlet*, act 1, scene 1:

> and the moist star
> Upon whose influence Neptune's empire stands
> Was sick almost to doomsday with eclipse.

———

Here the 'moist star' is the moon, so called because it controlled the tides. Neptune was the Roman god of the sea. Since Shakespeare and his contemporaries were aware of the mysterious connection between the moon and the tides, it is not surprising that *epidemics* should be blamed on mysterious influences. In 1743, when an epidemic broke out in Italy and spread across Europe, it was described in the London press by the Italian word for influence, *influenza*, and the name stuck.

FORMULA: the diminutive of **forma**: (L.) 'shape, form, nature, kind'. The word has a long history and has been applied in many situations even in recent times, when it can refer to a baby's drink or a class of racing car. The debate over formulae for chemical elements and compounds extended over many years. One author in 1789 depicted the formula for water as I-D, where I stood for Cavendish's inflammable gas and D for Priestley's dephlogisticated air. Eventually, element names were agreed upon, as well as their abbreviated symbols and the number of atoms in a molecule indicated by a number, written as a subscript. As organic chemistry progressed, it became clear that compounds existed that had the same composition but different chemical properties. This led Berzelius in 1830 to define these as *isomers* that differ in the internal arrangement of their atoms. In 1865, Kekule published the hexagonal structure of *benzene*, and in 1874, Van 't Hoff and Le Bel independently suggested that carbon chemistry needed to be considered in three dimensions arising from the *tetrahedral* bonding of the carbon atom. See also the entry for **CHIRALITY**. For derivation of *tetrahedral*, see chapter 4. The term *stereochemistry* was introduced by Meyer in 1890.

The **forma** root also occurs in *conformation*, which in biochemistry refers to the 3D shape of molecules, particularly proteins, where binding of a *ligand* may result in a *conformational change*. For examples, see the entry for **RECEPTORS**. In such cases, binding of an *extracellular* ligand to a *transmembrane receptor protein* results in a conformational change in the receptor that in turn triggers an *intracellular response.*

FRAGMENT: fragmentum (L.) from **frangere, fregi, fractum** 'to break'. The basic idea behind sequencing either a protein or a DNA molecule is to generate a series of *overlapping fragments* and, by sequencing these and ordering the fragments, derive the sequence of the original molecule. The same approach is used in *synthetic biology*, where in order to construct for example a synthetic chromosome, a series of overlapping fragments are synthesized and then stitched together in such a way as to eliminate overlaps. See under entry for MYCOPLASMA. The **frangere** root has also given rise to *fraction* and *fractional distillation*, *fractionate*, *fracture*, *refraction*, and *X-ray diffraction*.

FREE ENERGY: **en + ergon** 'work'. The change in free energy that results from a chemical reaction, represented as ΔG, is a guide as to whether a reaction can occur spontaneously or whether energy has to be supplied in order to make the reaction proceed. Another way of looking at the ΔG of a reaction is as a measure of the work (i.e. chemical work) that is provided by the reaction. For biochemical reactions that proceed under particular conditions of pH and temperature, modified values of free energy, $\Delta G^{0\prime}$, are determined.[44] Thus, the free energy of hydrolysis of the high-energy compound ATP is quoted as -7.3 kcal/mole, cf. for glucose 6-phosphate, it is -3.3 kcal/mole. The bond that is broken when ATP goes to ADP + Pi is sometimes described as a *high-energy bond*. This is not strictly correct because the $\Delta G^{0\prime}$ value given above is the difference in $\Delta G^{0\prime}$ between reactants and products as a whole. Synthetic or *anabolic pathways* are energy-requiring. This can be taken care of by the use of a *high-energy compound*, typically ATP, as a co-substrate. Alternatively, reactions with a $+\Delta G^{0\prime}$ may be driven by coupling with those that have negative $\Delta G^{0\prime}$ values.

If $\Delta G^{0\prime}$ for a reaction is negative, this means that it *can* occur but whether it *will* depends on the *activation energy* for the reaction (mixing hydrogen gas with oxygen gas does not produce water in the absence of a spark). While industry uses conditions such as high temperature and pressure to overcome activation energy barriers, living systems rely on enzymes, which reduce activation energies as a result of energy-yielding reactions involved in the binding of *substrate* to *enzyme*. See also under **ENZYME** heading.

GENOMES[45]—see also **CANCER GENOMES**.
THE HUMAN GENOME consists of twenty-two paired chromosomes together with the X and Y sex chromosomes. A first draft of its nucleotide sequence, close to 3×10^9 base pairs, was reported in 2001 and a refined version in 2003. This was the result of an international collaboration in which individual chromosomes sourced from different individuals were

sequenced in different labs. Thus, what is referred to as the *reference sequence* is a composite. This sequence still has gaps due to sequences that are refractory to sequencing methods. The availability of the human genome sequence has provided a major stimulus for medical research as well as providing insights into human evolution. The number of encoded proteins is about 20,000. For an average size of 500 amino acids, coded for by 1,500 nucleotides, it follows that only about 1 per cent of the genome codes for protein. A large number of functional non-protein-coding sequences, mainly involved in regulation of gene expression, are also being discovered, with the total number of nucleotides involved probably well in excess of those coding for protein.[46]

THE X CHROMOSOME. Sex determination in mammals depends on the X and Y chromosomes. The somatic cells of females contain two X chromosomes, those of males an X and a Y. It is believed that the X and Y chromosomes originated a few hundred million years ago from the same ancestral chromosome. The X chromosome consists of 153 million base pairs and codes for about 850 proteins. In order to obtain equal expression of genes on the X chromosome in both male and female cells, one of the X chromosomes in females is randomly selected for inactivation. This is referred to as *gene dosage compensation*. It follows that female tissues consist of two mixed populations of cells, each with either the maternally or paternally inherited X marked for inactivation. The two X chromosomes are designated Xa (active) and Xi (inactive). *X chromosome inactivation* (XCI) occurs as a result of synthesis of a long non-coding RNA, coded for on the Xi chromosome, and which proceeds to coat a large part of the Xi chromosome, which is silenced and becomes *heterochromatic*. Some genes on Xi escape inactivation, and the extent varies between species. In humans, up to a quarter of X genes escape. Many of the genes which do escape XCI are located within regions of the X referred to as being *pseudoautosomal*, these being genes which are also present on the Y chromosome, in this respect resembling the situation on other pairs of chromosomes, i.e. the *autosomes*.

THE Y CHROMOSOME. The smallest chromosome is the Y chromosome, consisting of 57 million bps but containing barely 200 genes and a high content of repeated sequences or junk DNA which made it difficult to sequence. This encouraged the rumour that the Y chromosome, together with half the world's population, in evolutionary terms, was on the way out. The argument went that because the Y chromosome only ever occurs as a single copy, there is no mechanism for repairing damage which inevitably must accumulate. This is in contrast to all the paired chromosomes, where genetic information lost from one chromosome can be restored by copying that information from the other one. Further analysis of the Y chromosome in the region containing its genes revealed that these *do* come in pairs. Most of these duplicated genes are arranged in eight *palindromes* within which each gene has an identical mirror image match. These palindromes are up to three million base pairs long. The diagram below illustrates this arrangement. Beginning at $(X)_n$ both copies when read 5' $\rightarrow$ 3' are the same.

$$\rightarrow \text{COPY 1}$$

5' X X A T G G C A T C G T T $(X)_n$ A A C G A T G C C A T X X 3'
3' X X T A C C G T A G C A A $(X)_n$ T T G C T A C G G T A X X 5'
COPY 2 $\leftarrow$

RETROTRANSPOSONS: **retro** 'backward' + **transpositus** (L.): the *ppp* of **transponere** 'to transfer'. Retrotransposons are genetic elements that are able to copy themselves and spread by a *reverse transcriptase*-mediated mechanism. These are DNA sequences which, when transcribed into RNA, are then converted to double-stranded DNA copies by reverse transcriptase. These copies can then randomly reinsert into the host genome at new sites.

Retrotransposons are classified into two groups: LINES (long interspersed nuclear elements) and SINES (short interspersed nuclear elements). Most LINES are 6–8 kb in length and are characterized by long terminal repeats, LTRs, which are also a feature of retroviruses. It appears that retroviral infection of germ-line cells is the origin of LINES containing LTRs and which encode the enzyme reverse transcriptase. The human genome contains about 500,000 copies of the L1 LINE retrotransposon, equivalent to 17 per cent of the genome. Only about 100–150 of these are evolutionarily recent and lack mutations that would affect their spread in somatic tissue. Various host defence mechanisms that silence retrotransposons have also evolved. Also present in the human genome are ancient non-LTR LINES.

The second class of retrotransposon, SINES, are present in about 1.5 million copies and comprise 11 per cent of the human genome. SINES derive from small RNA molecules, including *transfer RNAs*, and have arisen as a result of the conversion of RNA into double-stranded DNA by the reverse transcriptase expressed by LINE elements. The possible insertion of SINES and LINES into coding or regulatory sequences presents obvious mutagenic dangers. More than 100 instances of human disease have been attributed to retrotransposon insertions, because of both L1 and the common SINES Alu sequence.[47]

The total readily recognizable retrotransposons account for 35 per cent of the human genome. More detailed searches for repeated sequences have been carried out based on the idea that retrotransposons have almost certainly been accumulating for hundreds of millions of years and at the same time have been much altered by *mutation*. This approach suggests that these types of sequences may account for up to two thirds of the human genome.[48]

Another 2–3 per cent of the human genome is accounted for by *transposons*. These are DNA sequences that encode a *transposase*, an enzyme that enables movement and/or proliferation of the transposon. Other types of sequence in the human genome are the *simple sequence repeats* characteristic of heterochromatin together with separate examples of simple sequence known as *variable number tandem repeats* (VNTRs), which on account of their variability find *forensic* applications. VNTRs are estimated to account for 3 per cent of the human genome. There is still much of the human genome that is unaccounted for. This is sometimes referred to as the dark matter of the genome, but more often as junk DNA.

HUMAN GENOME VARIATION. *Next-generation rapid sequencing* has enabled a large-scale investigation of the extent of human genome variation. In the work quoted in the reference below, complete genome sequences were obtained for 2,504 individuals from twenty-six different populations and these sequences compared to the original reference sequence.[49]

In this study the typical genome differed from the *reference genome* at 4.1 million to 5.0 million different sites. Although >99.9 per cent of variants consist of *single-nucleotide polymorphisms* (SNPs) and short *indels* (insertions and deletions), structural variants defined as involving fifty or more bps affect more nucleotides. The typical genome contains an estimated 2,100 to 2,500 structural variants (~1,000 large deletions, ~160 copy-number variants, ~915 SINE insertions, ~128 LINE insertions, ~51 composite insertions, and ~10 inversions). Altogether about twenty million bps of sequence are affected. While most common variants are shared across the world, rarer variants are restricted to closely related populations. The typical genome contains 149–182 sites with protein-truncating variants (*frame shift mutations*), 10,000 to 12,000 sites with protein-sequence-altering variants. In addition, based on the occurrence of homozygous deletions, there are 240 genes apparently dispensable. This set is enriched for *immunoglobulin domains*, possibly reflecting individual differences in disease susceptibility. Between 459,000 and 565,000 variant sites overlap known regulatory regions, which probably goes a long way to explaining human variation at the phenotypic level. Most human genome variation occurs in Africa, consistent with the evidence that humans originated there and that only a sample of African genome variation is present in non-African populations.

GENOMICS

WHOLE GENOME DUPLICATION. It had been speculated that whole genome duplications (WGDs) had provided a means by which simple organisms evolved into more complex ones. Early anatomists noted that, unlike other invertebrates, the primitive *tunicates* (phylum *Chordata*, subphylum *Urochordata*) and *lancelets* (phylum *Chordata*, subphylum *Cephalocordata*) share key anatomical features with vertebrates and were regarded as forerunners of vertebrates. Analysis of features of the genome sequence of a tunicate, *Ciona intestinalis*, with those of the vertebrates puffer fish, mouse, and humans clearly indicated that two WGDs of a primitive chordate genome had set the scene for vertebrate evolution.[50]

When it comes to fish, a third round of WGD, the teleost-specific duplication, occurred approximately 300 million years ago, and a fourth round occurred at about 100 million years ago that gave rise to the ancestor of the rainbow trout. This last event is of interest because it provides an opportunity to investigate the results of a relatively recent duplication.

___

151

Here the two ancestral *subgenomes* have remained *colinear*, while about half of the protein-coding genes have been lost, mainly after conversion into *pseudogenes*, i.e. genes that are no longer functional because of mutation. The conclusion here is that WGDs are followed by slow alteration of the genome and not by a massive and rapid reorganization.[51]

Comparison of the genome of the model organism *Saccharomyces cerevisiae*, a single-cell yeast, with those of its relatives indicates that it is also descended from an ancestor that underwent WGD. There is evidence that *Arabidopsis thaliana*, the plant *model organism*, is also the product of WGD.

HUMAN PREHISTORY. While genomic studies as described above have contributed to our understanding of evolutionary events as far back as several hundred million years, genomics, together with advanced techniques for the extraction of ancient DNA, is also shedding light on numerous aspects of human prehistory. Fossil evidence indicates the emergence of anatomically modern humans in southern, eastern, and central Africa in the region of 350,000 to 260,000 years ago (350–260 kya), and further fossil evidence for anatomically modern humans in Ethiopia is dated to 150–190 kya. Human dispersal out of Africa is dated to 50–100 kya, and the data so far is consistent with a single dispersal. Related *hominins*, the *Neanderthals*, left Africa about 400 kya and, according to the fossil record, became extinct at 40 kya.

Genomic sequences have been determined for Neanderthals and another extinct hominin, the *Denisovans*. Why the Neanderthals and Denisovans died out is not known, but what is known from the genomic data is that humans interbred with both these archaic hominins. All non-African populations contain about 2 per cent Neanderthal DNA. The current evidence is consistent with a single admixture of Neanderthal DNA that occurred about 60 kya.

The archaeological evidence is that humans have occupied Australia for 55–60 kya. During the initial occupation Tasmania, the Australian mainland and Papua-New Guinea were a single land mass known as Sahul. Initial genomic studies on Aboriginal Australians and Papuans indicate they had separated from Eurasians by 51–72 kya, while separation of Australians from the Papuans occurred about 37 kya. As well as 2 per cent

Neanderthal DNA, there was an admixture of about 4 per cent Denisovan DNA into the Sahul population, estimated to have occurred 44 kya. The genomic data is consistent with just a single founding event in Sahul.

Similar genomic studies are being used to document the arrival of humans in America. All the evidence points to a people related to indigenous Siberians being the first to cross into America. Archaeological results point to their arrival around 12–14 kya. A significantly earlier date is regarded as unlikely but not impossible because much of North America was until then covered with a large ice sheet. However, recent investigation of a site in Mexico indicates it was occupied 30 kya, suggesting that the earliest inhabitants may have followed the coastline south to avoid the ice.

Genomic analysis of ancient remains has shown that the invention of agriculture in the Fertile Crescent about 11 kya was followed by the migration of populations from around this area into Europe, which resulted in the displacement of the indigenous hunter-gatherers. These analyses show that the introduction of agriculture to Europe involved the spread of people and not simply ideas.[52]

GENOME-WIDE ASSOCIATION STUDIES (GWAS). Following determination of the human genome, it was hoped that it would be possible to correlate particular sequence variants with the occurrence or at least the predisposition or susceptibility to particular diseases. By means of GWAS, it proved possible to show that the occurrence of many diseases does correlate with the presence within the genome of particular examples of *genetic variants*, usually *SNPs* or *indels*, located at known positions in the genome. In order to make statistically significant correlations and hence to identify the genes involved, GWAS analyses need to be applied to several thousand patients (the more the better) and a similar number of controls, each tested against a set of about 500,000 variants (the more the better) and, as near as possible, evenly distributed along the genome. The GWAS catalogue as of July 2017 consisted of 3,055 publications and 44,619 *variant-trait associations*.[53]

For most diseases, multiple loci have been implicated. A problem that became apparent was that in most cases, these correlations did not account for the known *heritability* of these conditions.[54] The likely reason is that many diseases or traits are the result of very many minor genetic contributions and these will only be detectable by using larger numbers of patients.

THE UK BIOBANK. This consists of a prospective cohort study with 500,000 volunteer participants, aged 40–69 when recruited during 2006–2010. All of these provided data on socio-demographic, family history, lifestyle, and health-related factors and completed what amounts to a comprehensive medical investigation. They provided blood, urine, and saliva samples and were evaluated for known biomarkers of disease and have been genotyped for subsequent GWAS investigations using genetic variants at 800,000 sites. In this project, the collection of detailed clinical and biological information is equally as important as the genetic component in order to obtain a combined comprehensive explanation of the genetic, environmental, and lifestyle influences on human disease. This approach is already unravelling both genome- and phenome-wide disease associations and represents a major innovation in modern medical research. The data obtained here is available to any health researcher on an open-access basis.[55]

THE EARTH BIOGENOME PROJECT. The proposal here is to decode the DNA of every plant and animal on earth. It is asking for US $4.7 billion to sequence all 1.35 million known eukaryotic species on earth. At present, there are 3,300 such genomes in US National Center for Biotechnology Information (NCBI) or 0.2 per cent of the total. A twofold justification is given for such a project—one economic, one environmental: possible discoveries leading to new biofuels, drugs, and useful agricultural traits; preserving biodiversity—the earth's sixth great extinction event is firmly underway with 50 per cent of current biodiversity possibly lost by the end of the century.[56]

CRISPR and GENE EDITING. CRISPR, or *clustered regularly interspersed short palindromic repeats*, refers to features of DNA sequences found in bacteria which are part of an *adaptive immune system* which bacteria use to protect themselves from infection by foreign DNA in the form of bacterial viruses (i.e. *bacteriophages*) and *plasmids*, which consist of transmissible DNAs also capable of *autonomous replication*. In recent years, the term CRISPR has come to denote an extraordinarily useful technique for the *specific* modification of eukaryotic genes. A brief description of how CRISPR operates in bacteria helps in explaining its application to eukaryotic gene editing.

The short palindromic repeats that occur in bacterial DNAs are fragments of foreign DNA, relics of previous infections. These are transcribed into short RNA molecules (known as *guide RNAs*) that associate with an endonuclease capable of producing double-strand breaks in DNA. This enzyme is referred to as *cas*, for CRISPR-associated. Entry of foreign DNA into the cell attracts the attention of the RNA-cas complex. If the incoming DNA contains a sequence corresponding to that of a guide RNA and this implies formation of a transient *DNA-RNA hybrid*, the endonuclease is activated and cleaves the foreign DNA.

So how is the CRISPR system applied to gene editing in eukaryotic cells? Obviously, it is necessary to know the sequence of the gene to be edited; also, the site within the gene to be edited needs to be exactly defined, and in normal circumstances, the experiment will be carried out on cells growing in tissue culture. It is necessary then to introduce into the cell a guide RNA ~30 nucleotides long which spans the site to be edited, together with the cas endonuclease. Both of these are introduced as *plasmids* which have been engineered to produce these two components inside the cell. Following cleavage of the specified gene, the *double-strand break* will be repaired by the *non-homologous end-joining* (*NHEJ*) system that in most cases carries out error-prone repair.

The aims of gene editing experiments will vary—from simply inactivating a gene in order to study the effect on phenotype, the reversal of a natural mutation or insertion of a sequence that may 'improve' the gene function or prevent disease. Because different types of bacteria have developed different versions of the CRISPR-cas system, a particular system may be preferred, depending on the aim of the experiment. The reference by Jinek et al. (2013) is an early paper describing the CRISPR technique; those by Cohen (2019) describe some of its varied applications.[57]

GLUCAGON and GLYCOGEN: glukus + ago, actus (L.): 'to do, make': i.e. to make glucose—from glycogen. Glucagon is a 29 amino acid peptide hormone synthesized in the α-cells of the *islets of Langerhans* in the *pancreas*. Its effect is to promote the breakdown of glycogen to form

———

glucose in response to low levels of glucose in the blood. Its action is therefore the opposite to that of *insulin*. Glucagon functions by binding to G protein-coupled receptors (GPCRs) located on the surface of liver cells and, to a lesser extent, muscle cells. This leads to activation of intracellular *adenyl cyclase* and the synthesis of *cyclic AMP*, which ultimately activates the enzyme *phosphorylase*, which degrades glycogen to glucose-1-phosphate. In line with one of the major functions of the liver being to maintain blood glucose levels, glucose-1-phosphate is converted to glucose-6-phosphate by phosphoglucomutase and then to glucose by glucose-6-phosphatase. *Adrenalin* also causes glycogen breakdown, acting mainly on muscle rather than liver, and acts via the *second messenger* cyclic AMP. Derivation: both glucagon and glycogen contain the Greek root **glukus** meaning 'sweet'; *-agon* comes from **agein** meaning variously 'to lead, bear, stimulate', cf. *agonist*, while the *-gen* suffix of glycogen indicates its ability to generate glucose, cf. hydrogen, halogen, etc.

GLYCOLYSIS: glukus- + **lusis** 'loosening' and **GLUCONEOGENESIS: glukus** + **neo** 'new' + **genesis** 'producing'. Glycolysis is an ancient metabolic pathway employed by a vast range of organisms. It consists of the multistep enzymic conversion of a 6-carbon sugar, *glucose*, into two molecules of a 3-carbon compound, *pyruvate*, with the release of energy. Gluconeogenesis is the synthesis of glucose from pyruvate. It is estimated that under conditions of starvation, there is sufficient glucose, either free or in the form of glycogen, to satisfy the requirements of the human brain for not much more than a day. In order, therefore, to supply glucose under these conditions, various other body constituents, such as amino acids, are converted to *phosphoenolpyruvate* and then to glucose by gluconeogenesis. Gluconeogenesis is not simply the reverse of glycolysis because glucose has to be activated by phosphorylation for glycolysis and likewise pyruvate for gluconeogenesis. These phosphorylation reactions are subject to stringent regulation of the activities of the enzymes concerned so that glucose is only metabolized or synthesized as required. Nonetheless, the steps between *fructose 1,6-bisphosphate* and phosphoenol pyruvate are common to both pathways and are fully reversible. For more on carbohydrate metabolism, go to https://en.wikipedia.org/wiki.carbohydrates.

GLYCOPROTEINS: Chains of sugar residues attached to proteins are termed *glycans*. Attachment to protein is either via *glycosidic* linkage to the hydroxyl groups of serine or threonine side chains or to the amide nitrogen of asparagine residues. These oligosaccharides are also referred to as *O-glycans* and *N-glycans* respectively, and typically consist of about twelve residues of various sugars. At least half of all mammalian proteins are *glycosylated*. The sugars can include glucose, galactose, and their N-acetyl derivatives, also mannose, fucose, and sialic acid. The N-glycans can either consist of a variety of these sugars or be composed mainly of mannose residues. Glycoproteins perform many different functions, and the function of their sugar components is likely to be equally varied. In many cases, it is known that they contribute to protein stability. In other situations, glycosylation occurs as the newly synthesized peptide chain emerges from the ribosome. Glycosylation may play a role in folding of the peptide chain. Glycoproteins in the plasma membrane invariably have the sugars positioned extracellularly. The distal sugar in these is usually sialic acid, and many differently modified sialic acids are known, suggesting that these may be part of a cell identity tag. The initial step in the binding of influenza virus to a susceptible cell involves the action of viral neuraminidase on sialic acid residues located on membrane oligosaccharides. See also PROTEOGLYCANS under **MUCO**- heading which consist of glycosaminoglycans attached to proteins and where the carbohydrate can account for up to 95 per cent by weight.

HELIUM: helios (Gk.) 'sun'. The name of this inert gas derives from the fact that it was first detected as a spectral line in sunlight in 1868, initially in the corona of a solar eclipse and later by Lockyer using a special telescope in normal sunlight. He called this new element helium, with the *-um* ending because he thought it was probably a metal. Helium was not found on earth until 1882, again detected by its spectral line during analysis of lava from Mount Vesuvius. Commercial quantities of helium are purified from natural gas, which can contain up to 7 per cent helium. A major use for helium is as liquid helium which, at -269°C, is used to cool the semiconductors which produce the magnetic fields used in magnetic resonance imaging (MRI). Lockyer;s other claim to fame is as the founder and first editor of the science magazine *Nature* in 1869.

HELIOCENTRIC: the model of the solar system proposed by Copernicus in 1543, in which the earth and the other planets orbit the sun and which, when generally accepted, displaced the previous *geocentric* model.

HELIOTROPISM: **helios** + **tropos** 'turning' is the tendency of plants and particularly flowers to point towards the sun.

HISTO-: **histos** (Gk.) 'web, tissue'. The subject *histology* consists of the study of tissue properties and composition. Various biological compounds are named accordingly, including the amino acid *histidine*, the *histones* (a group of basic proteins associated with DNA in chromatin), *histamine* (a compound released from tissues in response to injury and in response to allergic reactions which are treated with *antihistamines*).

HOMOLOGY: the property exhibited by protein and nucleic acid sequences, when aligned, of being recognizably related. Such sequences are described as *homologous*, and each sequence is a *homologue* of the other. A further distinction is made among homologous sequences: for example, the haemoglobins which occur in a wide range of different organisms and perform a similar function, i.e. the transport of oxygen in the blood, are clearly all descended from the same ancestral gene and are referred to as displaying ORTHOLOGY. Individual haemoglobins are *orthologous*. Another kind of homology is where a gene duplication has occurred and one of the gene products acquires a new function. An example is the duplication of the *lysozyme* gene that gave rise to *lactalbumin*, one of the subunits of the enzyme that synthesizes lactose. Lysozyme and lactalbumin are said to exhibit PARALOGY. As well as their amino acid sequences being 70 per cent homologous, they are also *paralogous*. Chemicals that belong to a series or are otherwise related are also referred to as homologues, e.g. *cysteine* and *homocysteine*, while a series of compounds such as *methane, ethane, propane,* etc. is referred to as a homologous series.

HYDROGEN—the water former. The example of hydrogen serves to illustrate why so much terminology in biological science, most of it invented in the last two hundred years, comes with a Greek or Latin stem. We have *hydroxides* and *hydrochloric* acid, *hydrocarbons* and *carbohydrates*, *hydrogenases* and *dehydrogenases*, the *hydrogen* bond and the *hydrogen* bomb. Virtually anything to do with water in a technical context also employs the **hudros** stem: the *hydronium* ion H_3O^+, *hydrolysate, hydrophilic, hydrophobic, hydrate, dehydrated, hydroponics, hydroelectricity*. Even the word *hydraulics* comes from **hudros** + **aulos** 'pipe', even though it is usually oil rather than water in the pipe. *Anhydrous* is applied to reagents to indicate purity or

158

that they lack water of crystallization or are suitable for reactions where water needs to be excluded. The Latin for water is **aqua**, and although there are many compound words in English based on aqua, only *aqueous* and *aquifer* come to mind as part of the scientific lexicon. One suspects that Lavoisier chose *hydrogen* rather than *aquagen* to make it clear that it was a chemical and not an alchemical word, or perhaps by this time, 'aqua' sounded a bit old-fashioned. *Aqua vitae*, literally 'water of life', was a term for alcohol, also known as ardent spirits (Latin: **ardere** 'to burn'). From the *OED*, 1586: 'From the lyes of wine is distilled a strong and burning aqua vita.' *Aqua fortis*, literally strong water, an early term for nitric acid—the first reference to it in the *OED* is from 1601. *Aqua regia*, 'royal water', a mixture of hydrochloric and nitric acids, was so called because it dissolved the 'noble metals' platinum and gold. The first reference to it in the *OED* is from 1594. Greek words related to **hudros** were **hidros** 'sweat' and **hygros** 'moist or wet'. A *hygrometer* is an instrument used to measure humidity, not to be confused with a *hydrometer*, used to measure the density of liquids.

HYPERSENSITIVITY refers to immune reactions occasioned by *allergies* (**allos** Gk. 'other' + **ergon** 'influence') where the response is excessive. The most common situations in which this occurs are *allergic rhinitis* (**rhis, rhin-** Gk. 'nose') or *hay fever*, *eczema* (**ek-** 'out' + **zein** Gk. 'to boil') and *asthma*, from the Greek **asthma**, from **azein** 'to breathe hard'. Usually there will have been previous exposure to the allergen. Reactions typically occur within minutes of exposure to the allergen. An IgE molecule specific for the allergen causes *mast cells* in *mucosal tissues* lining body surfaces, and *basophils* in the circulation, to release histamine and other agents (Edgar 2006).

The term *cytokine storm* is also used in the context of hypersensitivity. One explanation for why the 1918 flu pandemic was so lethal is that infected individuals were overwhelmed by a cytokine storm. *Cytokine*, from **kutos** (Gk.) 'vessel' + **kinein** 'to move' is a generic term that encompasses a large group of signalling molecules usually to do with immunity or inflammation. They are distinguished from *hormones* because they act locally, i.e. in a *paracrine* manner—see entry for **SECRETION MECHANISMS**. The term *cytokine* also includes the *interleukins*, named because they were originally shown to be secreted by *leukocytes*, i.e. white blood cells. For other words based on the **kinein** root, see under **KINASE**.

IMMUNITY. A biological arms race has gone on since soon after life emerged. Bacteria make antibiotics to kill other bacteria while bacteria, viruses, and other parasites ceaselessly probe animal defences. Microbes have the advantage of short life cycles and rapid evolution, but vertebrates have the advantage of an immune system that has been refined for many millions of years.

The terms *immune* and *immunity* derive from the Latin **munire, munivi, munitum** 'to fortify, strengthen, protect', the same root as in *munitions* and *ammunition*. The two main components of the immune system are the *adaptive* and the *innate*. The adaptive system, having encountered a pathogen once and mounted a response, has the ability to remember this if it meets the same pathogen again. This provides the basis for *immunization*, or *vaccination*, using vaccines containing either an *attenuated*, i.e. less virulent pathogen, from **ad- + tenuare** 'to make thin', or a purified component such as the protein coats or *capsids* of viruses such as flu, **capsa** (L.) 'case'. The term *vaccine*, **vacca** (L.) 'cow', derives from the pioneering use by Edward Jenner in 1796 of cowpox to provide protection against smallpox. He had noticed that people in contact with cattle infected with cowpox virus were less likely to contract smallpox. Another way of expressing this is that cowpox virus is sufficiently similar to smallpox virus, so *antigens* of the *benign* cowpox virus that present to the immune system generate *antibodies* able to bind to the corresponding antigens of the smallpox virus and thereby prevent its replication.[58]

The term ANTIGEN originally meant '*anti*body *gen*erator'. Traditionally, an antigen was a substance in a vaccine that was injected in order to generate an antibody. With the recognition of naturally occurring host antigens such as *cell surface antigens*, an antigen was redefined as that which reacts with an antibody. A more general, current definition is that an antigen is any material that, depending on circumstances, may give rise to an immune reaction. Material injected to generate an antibody is now referred to as an *immunogen*.

The immune system consists of two separate mechanisms: there is the *adaptive* system (**ad- + aptare** L. 'to adjust or modify') and the *innate* system (**innatus** L. 'innate, natural, inborn, inherent'). The advantage of the innate system is that it can react to an infection more or less immediately, unlike the adaptive immune system, which needs at least a week to respond. The fact that most people are not sick most of the time is probably due to the innate system. The beauty of the immune system as a whole is that it can take care of nearly every pathogen, the main exceptions being the more recently evolved eukaryotic parasites such as *Plasmodium falciparum*, which causes malaria and has learnt to evade host immunity by shuffling its surface antigens.

THE ADAPTIVE IMMUNE SYSTEM depends on two types of immune cell or *lymphocyte* (**lympha** L. 'clear water', such as spring water). The two main types of lymphocytes are known as B-cells and T-cells. The B-cells are mainly concentrated in the *spleen*, whereas the T-cells mature in the *thymus* and circulate in the *lymph system*. Both types are found in *lymph nodes*. The immature precursors of B-cells and T-cells along with the various other immune white blood cells are all produced in the bone marrow, derived from *haematopoietic* stem cells, from which all cellular blood components are derived. The adaptive immune system can itself be considered under two headings. One is the *humoral response* (**umor** L. 'liquid'), whereby B-cell-derived plasma cells produce soluble *antibodies* (*immunoglobulins*) that mainly circulate in the blood. The second type of adaptive response is known as *cell-mediated immunity*, in which a certain type of T-cell, the *cytotoxic* T-cell, recognizes and destroys cells infected by pathogens.

THE HUMORAL RESPONSE. When an antibody recognizes a microbial surface *antigen*, an *antigen-antibody complex* is formed that results in elimination of the microbe. Antibodies are large Y-shaped molecules, each containing four polypeptide chains, two heavy (H) and two light (L) chains. At the twin extremities of the Y, there are regions of amino acid sequence variability contributed by both H and L chains. These are the *antigen-binding regions* of the antibodies.[59] For details of antibody structure, go to https://en.wikipedia.org/wiki/antibody.

The extreme variability of antibodies results from a unique combinatorial re-assortment of gene segments in the antibody-producing *somatic cells* and from mutations in those sequences.[60] Each B cell synthesizes a single type of antibody. Only those cells which encounter a *non-self* antigen will be stimulated to proliferate and produce soluble antibodies. The unique antigen-binding specificity of the antibody produced by that B cell remains constant and is inherited by the daughters of that cell. This is the basis for *immunological memory*, i.e. the ability of B cells to respond to the presence of an antigen previously encountered years before and why vaccination can provide lifelong protection against many diseases. The total number of different antibodies is estimated to be of the order of 10^8.

Five different antibody classes or *isotypes* with different functions are produced by B cells: IgG, IgM, IgA, IgE, and IgD. Ig stands for *immunoglobulin*. The classes differ in their structure, including the types of both heavy and light chains. The predominant class in blood is IgG. Apart from binding to pathogens, IgG activates a protein cascade called the *complement pathway* (see below). IgE triggers *hypersensitivity* reactions by *activating mast cells*. IgA protects mucosal surfaces, such as in the stomach, lungs, and urinary tract. IgM, found mainly in the bloodstream, forms a large pentameric structure that efficiently sequesters foreign material.

CELL-MEDIATED IMMUNITY. The failure of initial organ transplants and skin grafts led to the realization that donor and recipient individuals chosen at random are genetically different in terms of their cell-surface antigens. This eventually led to the discovery of genetic loci known as the *major histocompatibility complex* (*MHC*). In humans, this is commonly referred to as the *human leucocyte antigen* (*HLA*) complex. HLA genes are classified as either class I, which are expressed in nearly all cells, or class II, where their expression is restricted to cell types known as *antigen-presenting cells* (APC) such as *macrophages* and *dendritic* cells, the name of the latter alluding to their irregular, ramifying shape (**dendros** Gk. 'tree'). Both these APC types are differentiated descendants of blood *monocytes*.

Each individual inherits three different class I HLA genes from each parent, a total of six. Class I genes are extremely *polymorphic*. Altogether there are several thousand *alleles* of these genes. This is why individuals selected at random are, in terms of their *cell-surface antigens*, genetically distinct from each other and why organ transplants from donors chosen at random suffer *rejection*. Each cell carries around 100,000 HLA proteins on its surface. They enable the immune system to identify and eliminate cells that have been infected by a pathogen.[61]

Normal turnover of cellular, i.e. 'self', proteins involves their degradation to peptides by an intracellular assembly termed the *proteasome*. If a cell is invaded by a virus, viral proteins synthesized in the course of viral replication will also be degraded to peptides. The HLA proteins bind those peptides, both self and non-self, and display those peptides on the cell surface. The critical step is that when peptide-bearing HLA proteins locate at the cell surface, only cells displaying non-self peptides are marked for destruction.

Recognition involves binding of the HLA-peptide complex to a receptor on a *cytotoxic T-cell*. Just as each B cell generates a unique antibody, each cytotoxic T-cell contains a unique receptor that is able to recognise a non-self peptide, and this is followed by activation of the T-cell. Activation causes secretion of cytokines and other proteins that induce *apoptosis* of the infected cell. Replication of the pathogen is prevented, and neighbouring cells are protected. How do uninfected cells displaying HLA proteins bearing self peptides remain unscathed? This is because T-cells that would recognize self peptides have been eliminated by negative selection in the thymus gland.[62] This is termed *immunological tolerance*. One reason the thymus gland was originally regarded as unimportant was because it was full of dead lymphocytes. Type II HLA proteins perform a similar function to that of class I, but the details are somewhat different. The antigen-presenting cells that express class II proteins are cells that engulf bacteria and degrade them. It appears that class I proteins deal mainly with viral pathogens while class II attend to bacteria and some viruses as well.

HLA GENES and TRANSPLANTATION. In spite of HLA heterogeneity, it is now possible in many cases to carry out successful organ transplants by optimizing the HLA match between donors and recipients. A worldwide database contains the HLA variants for eighteen million people. For an individual with a set of relatively common HLA variants, there might be 100–200 others with a similar set. However, nearly 6 per cent of people in this database are without any exact matches. *Immunosuppressant* drugs are now available that limit *rejection* of *organ transplants*. Continuing expansion of the HLA database together with use of immunosuppressants means that most patients can undergo successful transplantation.

Correlations are found between particular HLA variants and either susceptibility or resistance to disease. One example is an increased risk of developing multiple sclerosis. See entry for NARCOLEPSY under **SLEEP** for another example. In general, these correlations represent relatively low-level predispositions, and the mechanisms for these effects are not understood. HLA genes, like DNA variants in general, can shed light on the history of populations and their migrations, particularly when founding populations were small with a correspondingly limited amount of HLA variation.

INNATE IMMUNITY covers a variety of mechanisms. It is also referred to as *cell autonomous immunity* as distinct from *antibody-mediated immunity*. One aspect of the innate system is a heterogeneous group of white cells known as NK or *natural killer* cells. In contrast to cytotoxic T-cells, which eliminate infected cells that express HLA genes, NK cells seek out cells that do not express HLA genes. These can be cells infected by a pathogen that has contrived to prevent HLA gene expression (the biological arms race never stops). NK cells also target cancer cells that no longer express HLA genes because of mutation.[63]

The innate immune system consists of numerous systems that detect the presence of an invading pathogen. These are referred to as *sensors*, and they recognise *pathogen-associated molecular patterns* or *PAMPs*, i.e. signature components of pathogens. Once a sensor has detected a pathogen, it induces a *response* designed to eliminate the pathogen.

The presence of DNA in the cytoplasm of the cell can indicate the presence of a DNA-containing pathogen, a bacterium, or a virus, or it can be the result of cellular damage resulting in release of DNA from the nucleus or mitochondria. Either way, the cell needs to respond, and there are multiple systems that detect cytoplasmic DNA. An important, possibly pivotal, DNA sensor is the ribodinucleotide *cyclic guanosine monophosphate-adenosine monophosphate* (*cGAMP*) together with the enzyme responsible for synthesis of cGAMP, *cGAMP synthase*, abbreviated *cGAS*. cGAS, along with most, if not all DNA sensors, activates a cytoplasmic protein called *stimulator of interferon genes* (*STING*). STING is part of a *signalling cascade* that results in the production of antiviral *interferons* and cell-signalling proteins such as *cytokines*, as well as many other proteins whose roles in the innate immune response remain to be characterized.

COMPLEMENT is part of the innate immune system and consists of about twenty-six proteins that circulate in the bloodstream. In the late nineteenth century, Buchner observed a factor in blood serum that was able to kill bacteria. Soon afterwards, Paul Ehrlich found that antibodies induced by injection of an antigen were able to bind this factor. Ehrlich called the factor *complement* because of its apparent ability to complement the immune system. The proteins of the complement system consist of a series of proteolytic *proenzymes*, whose activation *cascade* is analogous to that involved in blood clotting. Several different activation pathways perform different functions. One such assembly of complement components, the *membrane attack complex*, kills bacteria by perforating their membranes. Another function is to stimulate *phagocytosis*, whereby components of activated complement serve as adaptors between antibodies bound to pathogens and receptors on the surface of phagocytic cells. Proteins functioning in this manner are referred to as *opsonins* and this method of stimulating phagocytosis as *opsonization*. According to the *OED*, the first recorded use of the word *opsonin* was in 1903, while George Bernard Shaw's play *The Doctor's Dilemma*, first produced in 1906, contains the line 'Opsonin is what you butter the disease germs with to make your white corpuscles eat them'. Shaw evidently was aware of the function as well as the derivation of opsonin: from **obsonium** (L.) 'that which is eaten with bread'.

INFLAMMATION from **flamma** (L.): *flame*. The term 'inflammatory response' is part and parcel of the immune response and is a blanket term for numerous separate responses that include expression of the genes for *cytokines* and *interferons* as well as transcriptional cellular responses that may include *autophagy*, *endocytosis*, *phagocytosis*, and *oxidative bursts* giving rise to *reactive oxygen species* or *ROS*—see under **ANTIOXIDANTS**. A key feature of the inflammatory response is the formation of a multiprotein complex known as the *inflammosome* in the cytosol of infected cells and in *macrophages*. The classical definition of inflammation, and it really is classical because it goes back to Celsus in the first century BCE, is *redness* (**rubor**), *heat* (**calor**), *swelling* (**tumor**), and *pain* (**dolor**), which could be a description of acute tonsillitis or a bee sting.

———

165

One of the first examples of *pathogen-associated molecular patterns or* PAMP detection and response to be investigated was the *lipopolysaccharide* (LPS) component of the cell wall of Gram-negative bacteria. Detection in this case triggers an inflammatory response that helps to eliminate the infection but can result in *anaphylaxis* (**ana** 'again' + **phylaxis** 'guarding' Gk.) or *toxic shock* if regulated improperly. For this reason, LPS is also known as *endotoxin*. Excessive responses of the immune system are not uncommon, as described above under **HYPERSENSITIVITY**.

Two further situations involving an inflammatory response can be distinguished. One example would be an infection by *rotavirus*, the double-stranded RNA virus that causes severe diarrhoea in children. Here the innate immunity sensor is part of an *inflammasome* that lyses infected intestinal cells. This lytic response minimizes viral replication and releases *cytokines* that protect neighbouring cells. In the normal course of events, good health is restored and is accompanied by reversal of the response. The other situation involves chronic diseases such as type II diabetes or problems to do with the cardiovascular system. This typically elicits a low-grade inflammatory response that is never resolved. The affected tissue is permanently host to macrophages, cytokines, etc., leading eventually to tissue damage.

INTEGER *n.*: in current English refers to a *whole* number, i.e. a number that is not a fraction. The same word in Latin means intact or entire: **in-** meaning 'not' while **teg-** is the stem of the verb **tegere, -xi, -ctum** 'to touch', so literally, untouched. The **teg-** stem also gives rise to *protect*, *detect* and *protégé*. INTEGRAL *adj.*: an essential or necessary part. Membrane proteins, such as receptors that completely traverse the membrane, are described as *integral membrane proteins*, i.e. they are part of the membrane.

INTEGRATE *v.*: from **integrare, -avi, -atum** 'to repair or renew', can refer to the assembly of components to make a complete object; also as in 'retroviruses integrate into host cell DNA'. DISINTEGRATE is self-explanatory. INTEGRINS are transmembrane receptors that mediate cell-cell interactions.

INTEGRITY *n.*: from **integritas** (L.), can refer to completeness or intactness: 'cell integrity was checked under the microscope'.

IRON: ferrum (L.), hence the symbol Fe and *ferrous* (Fe^{2+}) and *ferric* (Fe^{3+}) salts. The word in Greek was **sideros.** 'Iron is a precious cellular metal, sequestered by hosts and scavenged by pathogens.' As this quote indicates, iron is an essential requirement for all living organisms. The term *siderophore*, **sideros + phoros** 'bearing', refers to various molecules of microbial origin that are capable of chelating iron. Intracellular replication of the pathogen *Mycobacterium tuberculosis* relies on production of siderophores that scavenge host iron. Higher organisms rely on iron-containing *haemoglobin* for the transport of oxygen to tissues but, along with other aerobic organisms, also need iron for the *cytochromes* that mediate electron transport, while *non-haem iron* proteins include *ribonucleotide reductase* required for synthesis of DNA precursors.

As might be expected, *in vivo* levels of iron are tightly controlled. An excess of iron characterizes the disease known as *haemochromatosis*, a general term indicating excess iron in the body which can lead to tissue damage. The most common (but not the only) cause is an *autosomal recessive* mutation. Symptoms include tanning of the skin, hence the name of the condition, but more serious complications can also arise. A lack of iron usually manifests as *anaemia*. Humans have no means of excreting excess iron, so *iron homeostasis* depends on control of iron absorption by a hormone known as *hepcidin, hep-* because it is made in the liver, *-cidin* because it shows *microbicidal activity.* This activity is incidental to its role as the master regulator of iron metabolism. It is synthesized as a *preprohormone* (eighty-four amino acids), which is trimmed down to the active hormone containing twenty-five amino acids, including eight cysteine residues that form four disulphide bonds.[64] Hepcidin controls the amount of dietary iron entering the circulation by binding to and inactivating the iron channel *ferroportin* on the *basolateral* membrane of gut *enterocytes*. Ferroportin is a transmembrane protein which transfers iron from inside the enterocyte to the circulation. Hepcidin also controls ferroportin mobilization of iron in *macrophages*, *red blood cells*, and *hepatocytes* by a similar mechanism.

Transport of iron in the circulation is by the protein *transferrin*, which binds two ferric ions. The affinity of transferrin for iron is very high, quoted as 10^{23} M^{-1} at pH 7.4. Iron bound to transferrin is taken up into tissues by binding of transferrin to its receptor, followed by *receptor-mediated endocytosis*. Protons are then pumped into the resulting *endocytic vesicle*, and at a pH of around 5.5, the iron is released. The receptor with transferrin still bound returns to the cell surface.

The main mode of iron storage is as a *ferritin* complex in which up to 2,400 Fe^{3+} ions are stored as the oxide or hydroxide within a cage assembled from twenty-four ferritin protein subunits. Ferritin serves as a buffer against iron deficiency and iron overload.

ISOPRENE DERIVATIVES. CAROTENE: **carota** (L.) 'carrot'. A group of C_{40} polyunsaturated compounds generated by successive condensation of C_5 isoprenoid units. Formula for isoprene: CH_2=$C(CH_3)$-CH=CH_2. Carotenes are responsible for the orange colour of many fruits and vegetables, the colour due to the run of alternating or *conjugated* double bonds. Animals are unable to synthesize carotenoids. β-carotene, a symmetric molecule composed of two *retinyl* groups, is cleaved in the animal stomach to form *retinal*, a form of *vitamin A*. Vitamin A, in the form of *11-cis-retinal*, is attached covalently to *opsin* molecules in the rod cells of the retina. When this complex in the rod cells is activated by *photons*, the rod cells give rise to nerve impulses that stimulate the visual system of the brain. See entry for **OPSINS**. Carotenes also play a role in light harvesting during photosynthesis. *Coenzyme Q*, or ubiquinone/ubiquinol, which functions to carry electrons within the mitochondrial electron transport complex, consists of a *quinone* derivative joined to a 10-mer of the 5-carbon isoprene unit.

Naturally occurring hydrocarbons containing isoprene units are referred to as *terpenes*. Terpene itself is $(C_5H_8)_2$. Derivatives containing oxygen, such as *retinal*, are referred to as *terpenoids*. These occur in various essential oils and sustain the perfume industry. *Farnesyl pyrophosphate* is the C15 terpene intermediate in cholesterol synthesis. *Farnesylation* of proteins also occurs where it plays the same role as *myristoylation*—see under **FATTY ACIDS**. Farnesyl is the radical corresponding to *farnesol*, an essential oil used in perfumes and named from *Acacia farnesiana*, the sweet acacia tree, named in turn after a prominent Italian family. As mentioned under **ARCHAEA**, their membrane *lipid bilayers* consist mainly of isoprene-based molecules.

KARYO-: **karuon** (Gk.) 'nut or kernel', referring to the nucleus of the cell, at least to that of *eukaryotic* cells, *eu-* meaning 'true or normal'; bacteria and blue-green algae that have no nuclei are *prokaryotes*. Division of the cell nucleus during mitosis is referred to as *karyokinesis*, cf. *cytokinesis*, the division of the cytoplasm resulting in two daughter cells. The standard picture of a (resting) cell is one containing a single nucleus, but there are exceptions: muscle cells in large animals can be a metre or more in length and are *multinucleate*, containing up to 100 nuclei per cell, because of fusion of immature muscle *blast* cells. Some mature *hepatocytes* are *binucleate* and *polyploid* as well, containing eight, sixteen, or thirty-two times the haploid DNA content. The significance of this is not understood but is common to all mammals. The early *Drosophila* embryo is unusual in that it consists of a *syncytium*, **sun-** + **kutos** 'cell', with large numbers of nuclei produced by multiple rounds of DNA replication unaccompanied by normal cytokinesis. By about two hours into development, the syncytium contains around 1,500 nuclei, which cluster around its circumference. Formation of mononucleated cells then proceeds, producing the *blastoderm*.

KARYOTYPING refers to details of chromosomes made visible by arresting cell division using *colchicine* and then staining with one or more stains that produce a characteristic banding pattern. A karyotype may thus simply reveal the number of chromosomes characteristic of a species but for individuals can reveal details of abnormalities such as *chromosomal translocations*.

MEGAKARYOCYTE: **mega** 'large' + **karyo-** 'nucleus' + **cyte** 'cell'. See under **BLOOD**.

KINASE from the Greek verb **kinein** 'to move'. The equivalent Latin word is **movere, movi, motum**, hence *motion, motor neuron*, etc. Kinases are a class of enzymes (also termed *phosphotransferases*) that transfer a phosphate group from ATP to a wide range of acceptor molecules. The kinases most often referred to are the *protein kinases*. Determination of the human genome sequence revealed that it codes for over 500 protein kinases. Proteins can be phosphorylated at hydroxyl groups on serine or threonine residues and at tyrosine residues by *tyrosine kinases*, an important group of regulatory kinases. Phosphorylation in this fashion and dephosphorylation by specific *phosphatases* is a central feature of regulation in eukaryotic cells. Kinase enzymes specific for low-molecular-weight acceptors such as hexokinase and thymidine kinase are also prevalent. The term *kinetic*, as in *enzyme kinetics* and the *kinetic* theory of gases, derives from the corresponding adjective **kinetikos** 'related to motion'. **Kinema**, meaning 'motion', is where one goes to see moving pictures. **Kinesis**, the noun, also meaning motion, finds employment as *karyokinesis* and *cytokinesis*, referring respectively to division of the nucleus and division of the cytoplasm at the end of mitosis.

KINETOCHORE: **kineto- + khoros** 'place'. Kinetochores in higher organisms consist of up to 100 proteins that assemble at the *centromere* of each chromosome and provide attachment points for *microtubules* at the *mitotic spindle*. Further words based on the **kinein** root include locally acting hormones termed *cytokines* (see also under **HYPERSENSITIVITY**), *lymphokines* (i.e. cytokines secreted by lymphocytes, *chemokines* are cytokines with chemotactic activity while *angiokines* are secreted by endothelial cells lining blood vessels and *adipokines* are secreted by adipose tissue).

LIGATE: ligare (L.) 'to tie up, bind, bandage'. In the context of DNA metabolism, DNA fragments are *ligated* together, i.e. their ends are joined together by a *ligase* enzyme. In surgery, *ligation* refers to closing off blood vessels using a *ligature*. Related words: *ligament*: fibrous *connective tissue* that holds bones, cartilage, and joints together (the Latin word **ligamentum** meant bandage); *ligand*: a molecule that binds to another, typically to a transmembrane receptor. The *colligative* properties of solutions are the four properties that depend on the concentration of the solute, but not upon its identity: *freezing point depression, boiling point elevation, vapour pressure lowering,* and *osmotic pressure; obligation*: where one is legally or morally bound; *religion*: here it is people that are bound together.

LIPIDS: lipos Gk. 'fat' are more completely reduced than carbohydrates, and upon oxidation, they yield more than twice the energy obtainable from carbohydrates. They are non-polar in character, insoluble in water but soluble in *organic solvents* such as *chloroform* and *ether*. Lipids have three main functions: as *energy stores* in the form of *triglycerides*, as components of *membranes* in the form of *phospholipids* and *cholesterol*, while derivatives of both fatty acids and cholesterol function as *hormones* and *regulators*. *Triglycerides* consist of fatty acids attached via *ester linkages* to the three hydroxyls of *glycerol* and represent the major components of *animal fat* and *vegetable oils*. (Note that vegetable oils are chemically distinct from oils used to lubricate engines which consist of a mixture of *hydrocarbons*.) *Saponification* consists of the alkaline hydrolysis of fat or oil to yield soap, from the Latin for soap: **sapo-**, **sapon-**. See also **ADIPOSE TISSUE, CHOLESTEROL, FATTY ACIDS, LIPOSOMES** and **MEMBRANES**.

LIPOSOMES: **lipos** Gk. 'fat' + **soma** 'body'. These are vesicles bound by a lipid bilayer that can be produced by sonicating a solution of phospholipid. *Sonication* (**sonus** L. 'sound') consists of ultrasonic vibration. Liposomes have received attention as a means of *drug delivery*. The drug is included in the solution to be sonicated so that it will be included in the vesicle. Ideally, the surface of the liposome contains a specific binding molecule (or *ligand*) that will target the liposome to a *receptor* in a particular tissue. In an early version of this entry, I wrote 'This is an area where useful advances can be expected.' This prediction was spectacularly fulfilled when liposomes proved central to the development of mRNA vaccines.[65]

-LOGY, -LOGICAL, -LOGIST: these suffixes, tacked on to an appropriate noun, refer in turn to a discipline such as bio*logy*, the corresponding adjective as in bio*logical*, and a practitioner of said discipline as in bio*logist*. Similarly, we have zoology, histology, psychology, and a complete list of subjects identified by the *-logy* suffix would be a long one. These (as well as numerous other more or less related meanings such as *logarithm*) all derive from the Greek **-logia**, which comes from **logos** 'word', which had other context-dependent meanings. The Greeks themselves employed **-logia** to define subjects such as astrology and philology; **philologia** meant 'love of learning' in Greek, but *philology* means 'study of language' in English. According to the *OED*, a second group of English words ending in -logy or -logue derive from **legein** 'to speak'. These are words mainly concerned with aspects of speech and thought, such as *eulogy*, *apology, tautology, trilogy, ideology, homology, prologue, epilogue, dialogue*. New words, *neologisms*, are being coined every day. Logue and Conradi (2010) describe how a speech therapist cured King George VI of a stammer that made public speaking almost impossible for him. Curiously, the name of the speech therapist was Lionel Logue. There is a theory that people are drawn to occupations that are suggested by their names. Thus, William Wordsworth became a poet. The magazine *New Scientist* coined the term 'nominative determinism' to explain this. One of its readers claimed to have come across a paper on incontinence in the *Journal of Urology* by J. W. Splatt and D. Weedon.

The words *scientist, therapist* remind us that not all people who study or are expert in particular fields are described by words ending in **-logist**: practitioners of more ancient pursuits used suffixes such as *-loger* as in astrologer or *-logian* as in theologian. In biology, we have biochemists, botanists, and geneticists. Formation of the corresponding adjectives is straightforward using the suffix *-al* to give *biochemical, botanical,* etc. except, to my ear at least, *genetical* has never sounded quite right. This was in spite of the existence of a journal titled *Genetical Research*, which I see has, since 2007, mutated to *Genetics and Genomics Research*. The word *chemical* is also an example of what philologists refer to as back formation, which is where a noun gives rise to an adjective which in turn becomes a noun. So for chemistry, we have *chemical*, the adjective, as in chemical reaction and *chemical*, the noun, meaning a particular substance.

LYSIS: lusis (Gk.) 'loosening' from **luein** 'to loosen'; LYTIC comes from **lutikos** 'able to loosen'. Lysis describes cell destruction caused by rupture of the cell wall and/or cell membrane and mediated by mechanisms including osmotic effects, antibiotics, antibodies, enzymes, or detergent action.

LYSO-: prefix denoting agents able to cause cell lysis.

LYSOGEN: a bacterium carrying a *bacteriophage* integrated into the bacterial chromosome which, when activated to excise and replicate, results in lysis of the host cell. Resident bacteriophages are referred to as *prophages*; these are also known as *temperate* phages, as distinct from *virulent* phages, which do not integrate, and infection is followed by lysis.

LYSOSOMES: cellular *organelles* containing *acid hydrolase* enzymes that break down recycled and damaged cell material as well as invading bacteria and viruses which are first sequestered in membranous vesicles called *autophagosomes* which then fuse with lysosomes. The lysosome has traditionally been regarded as the garbage tin of the cell but more correctly should be seen as a cellular stomach since it recycles the products of macromolecular catabolism as well as being involved in plasma membrane repair, cell signalling, and under certain conditions, release of protein-degrading *cathepsins* that can lead to cell death by *apoptosis*. Lysosomes have attracted attention as the site of a large number of inherited diseases known as *lysosomal storage diseases*. A current estimate is that there are as many as seventy such human diseases, which display various symptoms. In general, these are the result of *mutations* in individual lysosomal enzymes. Another group of lysosomal enzymes are *targeted* in the membranous *organelle* known as the Golgi, for delivery to lysosomes by addition of mannose-6-phosphate. Mutation of the enzyme involved in this glycosylation results in these enzymes having no 'delivery address' and ending up being excreted in the urine.

-LYSIS when it occurs as a suffix can indicate lysis of a cell as in *haemolysis*, referring to the lysis of red blood cells or, more commonly, to a chemical reaction, e.g. the *hydrolysis* of an ester involves breakage of the ester bond with addition of the elements of water; similarly, the action of *phosphorylase* on glycogen in the presence of inorganic phosphate to produce glucose-1-phosphate is the *phosphorolytic* breakdown of glycogen, while the overall process of glycogen breakdown is referred to as *glycogenolysis*. *Glycolysis* is the series of enzyme-catalysed reactions that break down glucose, while *electrolysis* is the use of electricity to carry out reactions such as the formation of hydrogen and oxygen from water.

MEMBRANES: membrana (L.) 'parchment', contain three different types of lipids: *phospholipids, glycolipids*, and *cholesterol*. Phospholipids consist of glycerol, two fatty acids, and a phosphorylated alcohol. Phospholipids are also referred to as *phosphoglycerides*. Phosphorylation of the C3 position of glycerol already acylated at C1 and C2 is *phosphatidic acid*, which then forms an *ester linkage* with one of the following alcohols: *serine, ethanolamine, choline, glycerol,* or *inositol*. These phosphoglycerides are termed *phosphatidylserine, phosphatidylethanolamine*, etc.

Other phospholipids are based on *sphingosine*, formula: $H_3C(CH_2)_{12}CH=CHCHOHCH(NH_2)CH_2OH$. Various derivatives of sphingosine occur as constituents of membranes, particularly in nervous tissue. *Ceramides* (Gk. **keros** 'wax') consist of sphingosine joined in amide linkage via its amino group to a fatty acid. Further addition of *phosphorylcholine* to the terminal hydroxyl yields the phospholipid *sphingomyelin*. *Glycolipids* contain one or more carbohydrate residues attached to the terminal hydroxyl instead of phosphorylcholine. A single residue of either glucose or galactose generates a *cerebroside*, while an oligosaccharide in that position yields a *ganglioside*.

Sphingosine was named after the **Sphinx**, a Mediterranean deity of enigmatic character. An 1881 entry in the *OED*: 'A body remained insoluble which was of an alkaloidal nature, and to which, in commemoration of the many enigmas which it presented to the inquirer, I have given the name of Sphingosin'. The *OED* suggests Sphinx derives from the Greek **sphingein** meaning 'to bind tight', the connection here being that the Sphinx had the nasty habit of throttling, i.e. *asphyxiating*, people who failed to correctly answer her riddles.[66] **Sphingein**, appropriately, is also the root for *sphincter*. See also **PULSE**, chapter 4.

Phospholipids based on glycerol and those based on sphingosine have very similar overall architecture. In both cases, the hydrophobic hydrocarbon chains extend in one direction, while the negatively charged phosphate and the other hydrophilic groups extend in the opposite direction. Phospholipids are *amphipathic* molecules. The standard cell membrane model consists of a *phospholipid bilayer* such that the hydrophobic chains of the fatty acyl groups occupy a central non-aqueous environment while the *polar head groups* can interact with water on either side of the membrane.

Membrane fluidity varies in direct proportion to cholesterol content. Shorter fatty acid chain length has a similar effect as does fatty acid desaturation: the melting points of the C18 fatty acids oleic and stearic acid are 5—7°C and 69—70°C respectively. A further consequence of desaturation is that the *cis* double bond results in a kink in the fatty acid chain. These factors result in irregularity in membrane structure that is further increased by the presence of *integral membrane proteins*. See entry for **RECEPTORS**.

Archaeal membranes also consist of a lipid bilayer that performs the same function as in *Eubacteria* but the chemistry is quite different. In place of fatty acids, the lipid bilayer is composed of *polyisoprene* molecules joined to glycerol by *ether linkages*, and not all the polar head groups are the same as in *Eubacteria*.

For details of phospholipid and sphingolipid structures go to https://en.wikipedia.org/wiki/Phospholipid. Also https://.......wiki/Sphingolipid.

MENTOR: in academic circles, a mentor is someone (not necessarily the thesis supervisor) who advises, counsels, and generally keeps an eye on the progress of a postgraduate student. The word derives from when Odysseus, departing for the Trojan War, asked his friend Mentor to look after his son Telemachus. The *OED2* suggests the name was possibly chosen or invented as an appropriate one, given the related Latin word *monitor* has meanings that include 'guide' and 'teacher'. Burnside (2009) comments: 'Every busy person is called on to be mentor to someone just starting in the relevant field of endeavour.' He proceeds to wonder what the reciprocal is: 'If you are my mentor, what am I?' Neither *protégé* nor *pupil* gets a tick and certainly not *mentee*! He might have suggested *tormentor*, but he didn't. Recent usage suggests the answer to his question is indeed *mentee*.

METABOLISM comes from the Greek **metabole** meaning 'change', and it encompasses the full range of enzyme-catalysed reactions that are characteristic of living cells. Theodor Schwann (1810–1882), the eminent German physiologist, introduced the term. The root for metabolism, without the **meta-** prefix, is **ballein** 'to throw', with **bolos,** the noun, meaning 'a throw', which also shows up in words containing other Greek prefixes: *symbol, embolism, diabolical* as well as *hyperbola* and *parabola*. One of the meanings of **parabola** in Greek was 'parable', which evolved into Latin **parabolare** 'to speak' then into *parler* in French then *parley* and finally *parliament* in English. Metabolism can be broadly divided into ANABOLIC reactions, **ana-** meaning 'up', i.e. synthetic reactions, such as the assembly of macromolecules, which is energy requiring (bodybuilders take *anabolic steroids*) and CATABOLIC reactions, **cata-** meaning 'down', which are degradative reactions such as the breakdown of proteins, which may eventually end up as carbon dioxide and urea.

METABOLIC SYNDROME is a blanket term for a condition that is mainly characterized by obesity, high blood pressure, abnormal levels of blood lipids and insulin resistance. It carries an increased risk of cardiovascular disease and type 2 diabetes. The main causes are thought to be bad diet, in particular too much sugar, and a sedentary lifestyle with consequent lack of exercise.

METABOLOME refers to the complete spectrum of low-molecular-weight *metabolites* in a particular tissue in particular metabolic circumstances. It is a recently introduced term after the style of *genome*, *proteome*, etc. See also entry for **-OME**. This particular area of study is, naturally, METABOLOMICS.

ANTIMETABOLITES are compounds that interfere with cellular metabolism and are often *enzyme inhibitors*. Such inhibitors usually have structures similar to that of the normal substrate. The antibiotics *trimethoprim* and *sulphonamides* both interfere with bacterial folic acid metabolism. See also entry for ENZYME INHIBITION under **ENZYMES**.

METABOTROPIC: a term applied to receptors where binding of their ligands results in activation of a metabolic enzyme cascade such as glycogenolysis; cf. *ionotropic* receptors that regulate ion channels.

ANAPLEROTIC: **anaplerotikos** (Gk.) 'filling up' from **plethein** 'to be full' and **plere** (L.) 'to fill' and hence related to *plethora, complete, replete, deplete, supplement,* and *complement*; also *expletive*, which originally was a grammatical term referring to an extra word that 'filled out' a line of verse but which since the nineteenth century has acquired an additional meaning. Anaplerotic reactions are those which provide intermediates for a metabolic pathway. An example is the synthesis of *oxaloacetate* from *pyruvate* by *pyruvate carboxylase* that is activated by *acetyl CoA*, indicating there is a requirement for ATP generation by operation of the citric acid cycle and the electron transport chain. When ATP is abundant, oxaloacetate is withdrawn from the citric acid cycle and converted to *phosphoenolpyruvate* and eventually to *glucose*, a process known as *gluconeogenesis*, literally 'formation of new glucose'. Reactions which withdraw metabolic pathway intermediates for biosynthesis are termed CATAPLEROTIC reactions, examples being **GLUCONEOGENESIS**, also the conversion of oxaloacetate to the amino acid *aspartate*.

MEASURE: **metron** (Gk.) 'that by which anything is measured'. This has given us the *metric* system of measurement and *metre*, the standard unit of length. The Greeks had a few words involving measurement such as *geometry* and *symmetry*, but it is only since the beginning of modern science that anything that can be measured is done so using an instrument ending in -*meter*, such as *photometer, calorimeter, hygrometer*; also *stoichiometry* from **stoikheion** 'element', meaning the relative quantities of elements or compounds taking part in a chemical reaction. Then there is *econometrics, bibliometrics*, etc. Latin had the verb **metior, metiri, mensus** 'to measure', from which we get *mensuration*, the act of measurement; *immense*, the implication here being 'too big to measure'; also *commensurate* from **commensuratus**: 'equal' and *dimension*, being the measurable extent of something such as length, area, etc. **Metron** and **mensus** are related to the OE words *moon* (**men** in Gk.) and *month* (**mensis** in L.) both deriving from the Indo-European root **me** for moon, the moon in ancient times being used to measure time. *Menstrual* and *menstruation* clearly have a similar derivation. *Meniscus* is the diminutive of **men** and refers to the crescent shape of the new moon, which is similar in shape to the meniscus generated by the surface of a liquid in a narrow glass tube.

MILK

Through hir smokke, wrought with silk,
The flesh was seen, as whyt as milk.
Geoffrey Chaucer (1343–1400)

The cow is a creature of bovine ilk
One end moo; the other end milk.
Ogden Nash (1902–1971)

The Latin word for milk is **lac, lactis**, from which we get *lactation, lactose*, and in sour milk, *lactic acid* and its salt *lactate*; and *lactoferrin*, a milk protein which binds iron strongly, as a result of which it acts as an antimicrobial; also the bacterium *Lactobacillus*, some strains of which are used in the production of yoghurt. Two other milk proteins are *lactoglobulin* and *lactalbumin*; these two proteins are known as whey proteins since, along with other minor milk proteins, they remain in the *whey* fraction when the *casein* fraction, or *curd*, precipitates during cheese making. Milk is produced by *mammals*, named from **mamma** (L.) meaning 'breast' but also, according to Lewis and Short, a word used by Roman children to obtain their mothers' attention.

The Greek word for milk is **gala, galakt-**, hence the name *galactose* for the monosaccharide obtained from the disaccharide *lactose*; also *galactosidase* and *lactase*, enzymes that release galactose from a *galactoside* such as lactose. The medical profession always finds a Greek root hard to resist, so we also have *galactogogue*: a drug that promotes lactation (**agogos** 'leading') and *galactorrhea*: the spontaneous flow of milk (**rhein** 'to flow').

The word *galaxy* derives from the Greek word for what we call the Milky Way and which they called either **galaxias** 'milky one' or **kuklos galaktikos** 'milky circle'. According to the Greeks, the origin of the Milky Way was the result of an unfortunate galactorrheal incident when Hera, the wife of Zeus, was interrupted while breastfeeding. The English term Milky Way goes back to at least Chaucer and is a direct translation of **via lactea**, as it was known to the Romans. The Latin verb 'to milk' is **mulgere, mulsi, mulsum**. A related verb is **emulgere**, meaning 'to drain', the past participle of which has given rise to the English word *emulsion*, for which one dictionary definition is 'a liquid preparation of the colour and consistency of milk'. Burnside (2009) mentions that there is an English word *emulge* with the meaning 'to drain secretory organs of their contents' e.g. 'to milk out completely'. Its claim to fame is that it is one of only twelve words in the English language that end in *-lge*. These are *bilge*, *bulge*, *divulge*, and *indulge*, along with *emulge* and seven others which are about as commonly encountered as *emulge*. A Latin proverb was **hircum mulgere** 'to milk a buck', i.e. a he-goat, meaning 'to attempt the impossible'. *Milk* and ***mulg-*** both derive from the same ancient Indo-European stem, in line with recent evidence, outlined below, that milk and milk products such as cheese were being consumed not long after the agricultural revolution and the beginning of animal domestication.

DIET and EVOLUTION. Remarkably, it has been possible to obtain direct evidence that milk and milk products have been on the menu for humans for a very long time. Analysis of residues from independently dated pottery vessels from Anatolia in modern Turkey and surrounding areas reveals the presence of *fatty acids* derived from milk at least as far back as the seventh millennium BCE.[67]

Initially, this was surprising because it had originally been assumed that milk consumption following animal domestication would have been severely limited because of *lactose intolerance*. Utilization of the lactose of milk depends on the cleavage of lactose into its constituent monosaccharides, glucose and galactose, by the enzyme *lactase*. Infants are equipped to do this, but adult mammals do not normally produce lactase. For people who are lactose intolerant, the consumption of milk is not life-threatening, but the resultant nausea and diarrhoea ensure future avoidance of milk. A solution to this problem is cheese-making, where most of the lactose goes off in the whey while cheese has the added advantage of being able to be stored and transported. Archaeological evidence indicates that cheese was being made 8,000 to 9,000 years ago.

About 7,500 years ago, a mutation began to spread through Europe that resulted in *lactase persistence* throughout adulthood and therefore lactose *tolerance*. This lactase mutation is referred to as the *LP allele*. The fact that the mutation spread as rapidly as it did (more than 90 per cent of Europeans are now lactose tolerant) argues that it conferred a major selective advantage. The LP allele found in Europe, which is a *dominant mutation*, consists of a single nucleotide C → T transition, located over 19,000 bp upstream of the lactase gene *promoter*. The remote location of the mutation suggests the existence of a chromatin loop that brings the mutation adjacent to the *lactase promoter*. Meanwhile, about 70 per cent of the world's population is lactose intolerant although there are some other regions of the world, West Africa and the Middle East, where a high proportion of the population is lactose tolerant but these seem to be due to mutations distinct from the LP allele. The spread of the LP allele has also attracted the comment that this change in gene frequency, an example of *evolution*, depended on human behaviour in that ultimately it was animal domestication and the adoption of milk as an ingredient of diet that led to the spread of the LP allele and not the other way round.[68, 69]

In humans, the enzyme *amylase* that digests starch (**amylum** L.) is synthesized in both the *pancreas* and *salivary gland*. Studies in other animals indicate it was a duplication of the pancreatic gene that led to synthesis of *salivary amylase*. Comparison of human and Neanderthal genomes indicates that further duplication of the amylase gene occurred in the human lineage as the human diet became more starch-rich, particularly following the agricultural revolution. Different human populations have variable numbers of the salivary amylase gene with similar variation in the activity of the enzyme itself. It has also been reported that a correlation exists between the number of copies of the salivary amylase gene and the likelihood of a person becoming obese.

MITOCHONDRIA, mitos (Gk.) 'thread + **chondros** 'granule', are membranous structures or *organelles* located in the cytoplasm of the cell. It is now accepted that mitochondria arose as a result of a process known as *endosymbiosis*, in which a primitive precursor of eukaryotic cells was invaded by a bacterium containing an electron transport chain and able to carry out oxidative phosphorylation. Mitochondria are semi-autonomous organelles, containing DNA which, in humans, codes for a mere twenty-two tRNAs, two rRNAs, and thirteen proteins while 99 per cent of the proteins that function in mitochondria are encoded in the nucleus and imported from the cytosol. The main evidence that mitochondria and mitochondrial DNA are the remnants of a bacterial endosymbiosis event, crucial for *eukaryogenesis*, comes from DNA sequence comparisons of mitochondrial DNA from different eukaryotes whose mitochondrial DNAs encode different proteins derived from the invader. These DNA sequences point to a single endosymbiotic event involving invasion by a bacterium belonging to the *alphaproteobacterial* lineage.

It appears that most of the genes present in the original *symbiont* have either been translocated to the nucleus or discarded to different extents in different eukaryotic groups. Most eukaryotic cells contain mitochondria, and those eukaryotic species apparently lacking mitochondria nevertheless contain remnants thereof, suggesting that the symbiotic event occurred within an ancestor of all existing eukaryotic cells.

A characteristic feature of mitochondria is their *double membrane*. A membranous network in the interior, termed *cristae*, serves to increase the area of the inner membrane. The mitochondrial interior is referred to as the *matrix*. It is the inner membrane that accommodates the various complexes of the *electron transport chain* and the *ATP synthase* complex. Unlike the outer membrane, it is impermeable to all molecules apart from those with specific *entry/exit transporters* and which thereby provide metabolic connections between the cytoplasm and the *mitochondrial matrix*. These numerous *transporter* proteins, also called *carriers*, are embedded in the membrane, and each allows passage of a specific molecule.

The main function of mitochondria is to supply most of the energy requirements of the cell by carrying out *oxidative phosphorylation* whereby energy liberated by operation of the *citric acid cycle* and *fatty acid breakdown* is utilized to synthesize ATP. Electron transport in mitochondria involves the flow of electrons, derived from NADH and $FADH_2$, down a series of complexes terminating in reaction with oxygen to form water. At three different points, the energy of this stepwise oxidation is harvested by pumping protons across the inner membrane setting up a proton gradient. As the protons flow back, they do so through the enzyme complex *ATP synthase*. ATP synthesis occurs according to the following equation:

$$ADP^{3-} + HPO_4^{2-} + H^+ \rightarrow ATP^{4-} + H_2O$$

About thirty molecules of ATP are produced by the complete oxidation of each molecule of glucose.[70]

See under **ENZYMES** for websites showing NAD^+ and FAD structures.

MITOSIS and the CELL CYCLE. Mitosis is from **mitos** (Gk.) 'thread', from the threadlike appearance of chromosomes in the light microscope during mitosis. The *cell cycle* of a typical dividing diploid animal somatic cell is divided into the following stages or phases: *G1 phase*, where the G stands for 'gap' or 'growth' lasting possibly five hours. *S phase* corresponds to DNA synthesis, i.e. the complete replication of the genome, occupying around seven hours. *G2 phase* lasts say, three hours until the beginning of M phase. *M phase*, mitosis, one hour, in which cell division occurs. These times are indications of relative durations, but the overall length of the cell cycle can vary considerably. Note that the diploid cell in G1 contains 2n chromosomes, where n is the haploid number of chromosomes; this becomes 4n at the completion of S phase then back to 2n at the end of mitosis.

Duplication of each chromosome gives rise to a pair of *chromatids* that are held together by a complex of four proteins (termed *cohesin*) that binds at the chromosome *centromeres*. As mitosis begins, at prophase, the chromosomes condense and become visible. Transcription has ceased, the nuclear membrane disintegrates into small vesicles, and the *nucleolus*, the site of ribosomal RNA synthesis, also disappears. A kind of organelle called the *centrosome*, usually located near the nucleus, has also duplicated by this stage, and the two centrosomes migrate to opposite sides of the cell. *Microtubules* generated by the polymerization of *tubulin*, spread from the centrosomes to form the *mitotic spindle* and attach to *kinetochore* proteins at the centromeres, causing the chromosomes, now at metaphase, to align at the centre of the cell. During *anaphase*, the cohesins are cleaved, the microtubules contract, pulling the new *daughter chromosomes* to opposite sides of the cell. Division of the nucleus, or *karyokinesis*, then proceeds and is followed by *cytokinesis*, i.e. cell division.

Mitosis needs to be closely regulated. There exist a number of *checkpoints* that ensure that all chromosomes proceed through the various steps in unison. If this does not occur, then *aneuploidy* is the result, and *apoptosis* will be triggered. The regulated progression of a cell through the cell cycle occurs by way of a family of proteins termed *cyclins* that activate *cyclin-dependent kinases* or CDKs. Numerous cyclins and many CDKs are involved, and their concentrations vary with the stage of the cell cycle. See the PROTEIN DEGRADATION heading under **PROTEINS** for the *ubiquitin-proteasome* method of regulated protein removal.

MEIOSIS: **meioun** (Gk.) 'to lessen', a process in which diploid cells undergo one round of DNA replication followed by two cell divisions to generate *haploid gametes*. Gamete, meaning germ cell, derives from Gk.: **gamete:** *wife*; **gametes:** *husband*; **gamein:** *to marry* and **gamos:** *marriage*; cf. bigamy. Following a 2n to 4n replication as above, the initial steps of the first meiotic division are similar to mitosis whereby homologous chromosomes align at a spindle apparatus. At this stage, meiotic metaphase 1, recombination occurs between the chromatids of the homologous maternally and paternally derived chromosomes, a process termed *synapsis* from the Greek: **sun-** 'together' + **hapsis** 'joining' from **haptein** 'to join'. Further genetic mixing then occurs during anaphase as maternal and parental chromosomes randomly segregate to the opposite poles of the cell. The total number of different possible combinations of parental chromosomes is 2^n, where n is the number of chromosomes. For humans, the total number of ways of assembling chromosome pairs 1 to 23 by randomly mixing 23 maternal and 23 paternal is 2^{23} or about 8 million. The result of the second round of meiotic cell division is the formation of haploid gametes.

MONOCLONAL ANTIBODIES (**mono-** 'single' + **klon,** Gk. 'twig', 'branch'). *Multiple myeloma* is a type of cancer affecting antibody-producing B cells. As a cancerous B cell proliferates, the immunoglobulin fraction of blood may become dominated by the single antibody produced by the rogue B cell. Such antibodies proved useful in early work on antibody structure.

Multiple myeloma B cells have been exploited to obtain B cell lines that produce antibodies specific for individual antigens, in fact specific for a single *antigenic determinant* or *epitope*. These are termed monoclonal antibodies because they derive from a *single cloned B cell*. This involves using B cells that have responded to a particular immunogen and fusing these with multiple myeloma cells. Fused cells are known as *hybridomas*. This process effectively *immortalizes* the B cells. Unfused cells are eliminated by growth in selective medium with further selection for cells producing the required monoclonal antibody. Many such specific antibodies are in clinical use treating cancer and a variety of other conditions. These mainly have names ending in -*mab*.

MORPH- (Gk.): 'form' or 'character'.

AMORPHOUS: **amorphos** 'shapeless' from **a-** 'without' + **morphe**. MORPHOGEN: a diffusible signalling molecule which determines the pattern of gene expression and therefore of tissue development during embryogenesis. Another name for this process is *chemotaxis*. Morphogens act directly on cells in a concentration-dependent manner to induce or maintain expression of particular morphogenetic genes. Examples of morphogens are *epidermal growth factor* (EGF), *fibroblast growth factor* (FGF), and *retinoic acid*.

MORPHOLOGY: the study of the form and structure of organisms.

METAMORPHOSIS: **metamorphosis** (Gk.) 'transformation', an abrupt change in an animal's body structure with accompanying changes in habitat and behaviour. The occurrence of metamorphosis is familiar in *amphibia* and insects and is common in numerous other phyla. In amphibia, the larval stage such as the tadpole is confined to water and possesses gills, while the adult is an air-breathing form with lungs. An equally abrupt transformation occurs as the waterborne mosquito larva, the wriggler, becomes the adult mosquito while maggots turn into flies.

POLYMORPHISM: in biology it generally refers to multiple phenotypes within a single species, such as coat colour in cattle. Since all biological polymorphism is genetic, polymorphisms are apparent at the level of DNA as well. These may consist of *single nucleotide polymorphisms*, SNPs, or longer *insertions* or *deletion involving a few nucleotides* termed *indels*. Polymorphisms only persist if maintained by natural selection. Blood cells with nuclei having irregular lobular shapes are known as *polymorphonuclear leukocytes*.

MUCO- mucus (L.) 'nasal secretion'; **mucosus:** covered with or producing mucus. MUCOSA: abbreviation of **membrana mucosa** or mucous membrane: any of the moist membranes that line internal surfaces of the body and consist of a surface layer of epithelium containing *mucin-secreting cells*. MUCIN: a large group of high-molecular-weight *glycoproteins* that form viscous solutions and serve as lubricants as well as providing protection to internal body surfaces. Excessive production and secretion of mucins contribute to the formation of mucus that causes blockade of the airways in many lung diseases such as asthma. Mucins are produced by goblet cells and by mucous cells in submucosal glands. They are packaged in vesicles called secretory granules. Molecules that stimulate mucin secretion are called mucin *secretagogues*.

The importance of *dietary fibre* in the diet is starting to be understood as due to its role in keeping the mucous layer of the gut intact by encouraging the growth of good bacteria and preventing access to the gut lining by opportunist bacteria and toxins that cause *inflammatory responses*.

MUCOPOLYSACCHARIDES: also referred to as GLYCOSAMINOGLYCANS. These consist of repeating disaccharides containing an acetylated hexosamine joined to an *uronic acid* residue, which may also contain sulphate residues. Uronic acids are modified hexoses in which the $-CH_2OH$ at the 6 position is oxidized to $-COOH$, a carboxyl group. Examples of glycosaminoglycans are *chondroitin sulphate*, *keratan sulphate*, and *hyaluronic acid*. Since they are highly negatively charged, they have an extended conformation in solution and thus exhibit high viscosity. A number of inherited diseases, *mucopolysaccharidoses*, are known which are characterized by mutations which prevent breakdown of mucopolysaccharides. PROTEOGLYCANS consist of glycosaminoglycans attached to proteins in which the carbohydrate can account for up to 95 per cent by weight. These play important roles in connective tissue. See also entries for **CARTILAGE** and SUGAR DERIVATIVES under **CARBOHYDRATES**.

MUSCLE. The word *muscle* derives from the Greek root **mus, muos**, meaning muscle but which was also their word for mouse; the Romans picked up on this double meaning, which one feels must have started out as a joke: their word for mouse was **mus**, while the diminutive, **musculus**, which in the normal run of things would have meant 'small mouse', meant muscle. Meanwhile, the Linnaean *binomial* for the house mouse is *Mus musculus*. The *myo* root shows up in various components of muscle, including one of the major proteins involved in muscle contraction, *myosin*, while muscle *fibres* are composed of *myofibrils*. *Myoglobin* is a haem-containing, oxygen-binding protein that serves as an oxygen reserve for muscle. Other examples are *MyoD* and *myogenin*—two transcription factors specific to muscle cells or *myocytes*. Various diseases that affect muscle are referred to as *myopathies*; such muscles become *amyotrophic*, while muscle pain is *myalgia*. *Myasthenia gravis*, **mus** + **asthenes** 'weak', refers to conditions characterized by abnormal muscle weakness caused by autoimmune antibodies binding to the acetylcholine receptor. In addition to myosin, the other protein central to muscle action is *actin*, so named by Albert Szent-Györgi, a pioneer of muscle research, because of its ability to *activate* muscle fibre contraction *in vitro* in the presence of myosin and ATP.[71] The genes encoding myosin and actin are just two of more than 100 genes involved in the development of striated muscle in vertebrates. Different tissues have different actin and myosin *isoforms*.

Another Greek root commandeered to describe components of muscle is the word for 'flesh': **sarx, sark-**. The regularly repeating units along the axis of the myofibril are the *sarcomeres*, and it is the movement of the *actin* and *myosin* protein filaments of the sarcomeres, relative to each other, that results in muscle contraction. A muscle fibre corresponds to a single muscle cell, bounded by a cell membrane, the *sarcolemma*, while the cytoplasm within is referred to as the *sarcoplasm*. In muscle cells, it is the *sarcoplasmic* *reticulum* which corresponds to the endoplasmic reticulum in other cells. This is the membranous *organelle* that is a site of protein synthesis and in muscle also controls the release and reabsorption of Ca^{++} which, upon the receipt of a nerve impulse, triggers muscle contraction. Muscle cells in large animals can be a metre or more in length and are therefore among the largest cells that occur in animals, and a not unrelated feature, they are *multinucleate* and may contain up to 100 nuclei per muscle cell. The presence of multiple nuclei is the result of the fusion of immature muscle blast cells (*myoblasts*) to form the large mature muscle cell. For diagrams of a muscle fibril and fibre and the arrangement of muscle components, see Wikipedia at 'Muscle'.

Another example of the *sarc* stem is in *sarcosine* or N-methyl glycine. Formula: $CH3-(H_2N^+)-CH2-COO^-$, first isolated from muscle tissue and traditionally described as a breakdown product of creatine but now known to be synthesized, at least in some circumstances, by methylation of glycine by the enzyme glycine-N-methyl transferase. *Sarcopterygii-* (**ptero-** 'wing or fin'): a class of fleshy- or lobe-finned fish which include *coelacanths* and *lungfish*, regarded as the closest living ancestors of *tetrapods*, i.e. the terrestrial four-limbed vertebrates. *Sarcoma* is a type of tumour that develops in fibrous and connective tissues. A well-known example is the *Rous sarcoma virus* (RSV) isolated from chickens in 1911 by Peyton Rous, who showed that the tumour was transmissible when injected into other chickens. Further work on RSV, much later in the 1970s, led to the concept of the *oncogene* as a cause of cancer when inappropriately expressed. *Sarcophilus harrisii* is the name of the carnivorous marsupial better known as the Tasmanian devil. The *Sarcophilus* name alludes to their habitual fighting and biting. Its numbers have been much depleted by a facial tumour which, first seen in 1996, is also transmissible like RSV. Most bites occur around the mouth, and it is here that tumours develop and which, within a few months, cause death by starvation. See also XENOGRAFT under **XENO-**. *Sarcophagus*: **sarkophagos** 'flesh eating': in a case of linguistic evolution, sarcophagus came to mean a stone coffin, allegedly because a particular type of limestone was found to rapidly 'consume' the flesh of corpses laid within it. Then there is *sarcasm*, which comes from the Greek word meaning to 'tear flesh'—ouch!

The time before technology has been termed the *sarcocentric* era, when human and animal muscle power were the prime movers (Vogel 2001). The advent of the steam engine in the eighteenth century and its application to the operation of pumps and mills, jobs which until then horses had performed, made it necessary for the manufacturers of steam engines to indicate to potential customers the power of their steam engines in terms of the power that was obtainable from a horse, in other words, the horsepower (hp) of their engines. James Watt determined that a horse could walk an average of 2.5 times around a 24 ft. diameter mill wheel in 1 minute while exerting a force of 180 ft. lbs. Given that power = work/time = force × distance per unit time, this gave the result that 1 hp = 33,912 ft. lb. min^{-1}, which he rounded down to 33,000 ft. lb. min^{-1}. Watt's definition is therefore based on the power output that a horse can maintain over a working day but not its peak performance, which for large draft horses has been measured at between 10 and 15 hp. In modern parlance, 1 hp = 746 watts.[72]

Creatine and creatine phosphate together account for about 0.5 per cent of muscle weight. The name is based on **kreas,** another Greek word meaning 'flesh' or 'meat', which also shows up in *pancreas*. Creatine plays an important role in the supplying energy to the brain and particularly to muscle tissue in vertebrates. It is synthesized in the kidney and liver from parts of the amino acids glycine, arginine, and methionine. It is then transported in the blood to the brain and skeletal muscle, where it is converted by creatine kinase to creatine phosphate. Creatine phosphate, like ATP, is a high-energy compound. When energy demand is high, creatine kinase catalyses the conversion of ADP to ATP with creatine phosphate as the phosphate donor. It thus serves as a rapidly mobilizable reserve of high-energy phosphates in brain and skeletal muscle since the concentration of creatine phosphate in these tissues is a lot higher than that of ATP.

Estimates vary, but the amount of ATP present in muscle would keep a runner going for only one or a few seconds while creatine phosphate will support perhaps ten or twenty seconds of exertion. After that, it is muscle fat and glycogen that will satisfy requirements for perhaps an hour. The liver then has to take over. All this obviously depends on the level of exertion and the fitness of the runner. Fitness thus comes down to keeping up the supply of high-energy phosphate, i.e. ATP, and trained athletes have more and larger ATP-generating *mitochondria* than couch potatoes.

How often has muscle evolved? Until recently, it had been assumed that this had only occurred once, the indications being that it developed possibly 700 million years ago in simple eukaryotes where it carried out contractile functions to do with cell division or change of cell shape. A recent study searched the fully sequenced genomes of twenty-two animal species for *homologues* of forty-seven human muscle genes. This showed that some vertebrate muscle genes were present in animals that lack muscles, such as sponges. However, in a jellyfish (a *Cnidarian*), which has striated muscle, only one of the vertebrate striated genes could be detected, strongly suggesting that the presence of striated muscle in both vertebrates and jellyfish is the result of *convergent evolution*.[73]

MUSCLE as MEAT. A flesh-eating sarcophagus brings to mind the word *carnivore*, which also means 'flesh-eating', derived from **caro, carn-** 'flesh' + **vorare** (L.) 'to devour'. The animal kingdom can be divided into carnivores, *herbivores* (**herba** L. 'grass'), including koala bears, which are also *folivores* (**folium** L. 'leaf'), and *omnivores*, **omnis** (L.) 'all or every'. Obligate carnivores are referred to as *predators*, i.e. animals that prey on other animals, from Latin **praedator** 'plunderer, pillager'. Other words that incorporate the *carn-* root include CARNIVAL: **carn-** + **levare** (L.) 'to diminish or lessen', literally 'to cut back on flesh', referring to the long-standing custom in Catholic countries to eat, drink, and be merry at the Carnivale on Shrove Tuesday the day before the fasting period of Lent began on Ash Wednesday. A reasonable guess might have been that *carnival* derived from **carnem vale**, 'farewell to meat' as in the Latin version of 'goodbye': **ave atque vale** 'hail and farewell'. However, the *OED* prefers **carnem levare** on the basis of an early spelling 'carnelevale' that goes back as far as 1130. On reflection, it perhaps makes more sense to 'put the meat away', back into the corning liquid ready for Easter, rather than farewelling it as if becoming a vegetarian.

The compound CARNITINE plays a crucial role in the oxidation of fatty acids in mitochondria. Formula: $(CH_3)_3N^+\text{-}CH_2\text{-}CHOH\text{-}COO^-$. Long-chain fatty acids are activated by reaction with ATP and coenzyme A to form fatty acyl CoAs, which cannot traverse the *inner mitochondrial membrane.* Fatty acyl CoAs react with the hydroxyl group of carnitine to form *acyl carnitine.* Acylcarnitine is then shuttled across the inner mitochondrial membrane into the mitochondrial *matrix* by a membrane-bound specific *translocase.* Fatty acyl CoAs are then reformed, and carnitine is re-exported by the translocase.

Other words deriving from the **carn-** root include *carnage, carnal, incarnation,* and the flowering plant *carnation,* where the colour of the flowers resembles that of flesh, from **carnatio** (L.) 'fleshiness'.

MUTATION: mutare (L.) 'to change'. Also *mutate, mutant, mutagen.* A mutation is any change in a DNA sequence compared to a *reference* or *wild-type* sequence. Such changes can result from many different causes, including replication errors. The simplest mutational change is the substitution of one nucleotide for another. These are termed SNPs for *single-nucleotide polymorphisms.* Where a pyrimidine is replaced by another pyrimidine or a purine by a purine, this is classified as a *transition*; alternatively, when a pyrimidine is replaced by a purine or vice versa, this is a *transversion.* SNPs that occur within a protein coding sequence are classified as *synonymous* or *non-synonymous* depending on whether the mutation changes the amino acid coded for. If it does, then this is a *missense* mutation. It is also possible that a SNP will generate one of the three stop codons, then it is termed a *nonsense* mutation and by definition results in premature termination of translation. Another type of mutation is the *indel,* where one or more nucleotides are inserted or deleted. An indel involving one or two nucleotides alters the triplet *reading frame* and has the effect of completely scrambling the downstream encoded amino acid sequence. These are known as *frameshift* mutations. Indels involving multiples of three nucleotides either add or delete an amino acid and do not alter the downstream reading frame and *may* therefore still produce a functional protein.

ORIGINS OF MUTATIONS. Mutations can occur as a result of DNA *damage* which either goes unrepaired or is repaired inaccurately. Most mutations are the result of exposure to environmental mutagens. Some mutations occur because of the *intrinsic instability* of DNA nucleotides. One example is the *N-glycosidic bond* that joins purines to deoxyribose which can spontaneously hydrolyse, estimated to result in 5,000 to 10,000 *depurinations* per human cell per day. A mutation can result if this is not repaired correctly. Another example of DNA instability is *deamination* of cytosine which occurs about 100 times per cell per day. The product here is uracil, which is recognized by repair enzymes and removed. It is the occurrence of this reaction that likely explains why DNA contains thymine instead of uracil, the equivalent base that occurs in RNA. If uracil was present in DNA, there would be no way for the cell to distinguish normal U from U produced by deamination of C. But this is not the whole story for cytosine deamination. A common *epigenetic modification* of DNA is methylation of cytosine residues to produce 5mC. Deamination of 5mC produces thymine, which now becomes part of a G-T base pair. Upon replication, one daughter molecule will have the original G-C base pair, the other an A-T base pair. Most methylations of C occur when it is part of a CpG *dinucleotide*, which is the recognition sequence for the methylating enzyme. As a consequence over evolutionary periods of time, CG dinucleotides have been gradually replaced and occur in vertebrate DNA at only 10 per cent of the frequency expected, i.e. the frequency of GC dinucleotides. The depurination and deamination reactions outlined above are examples of reactions which are *endogenous* to the cell. Reactive oxygen species (ROS) that are generated intracellularly can also modify DNA and result in mutations if not repaired. Mutations also occur because of replication errors but, in normal circumstances, at a very low level, largely from a *proofreading mechanism* that detects and reverses the incorporation of an incorrect base. Estimates of the overall rate of mutation in human DNA are in the range 0.5 to 1.0×10^{-9} base pairs per person per year.[74]

REPAIR MECHANISMS exist for nearly every conceivable kind of DNA damage. Rather than responding to a particular type of nucleotide damage or modification, it is any distortion of the helix/nucleosome that is commonly recognized. A variety of enzymes specialized for repair include N-glycosylases for removing damaged bases, endo- and exonucleases, repair DNA polymerases, and DNA ligase. *Translesion synthesis* may occur in the situation where excessive damage has overwhelmed the repair machinery and the cell has two options: one is to trigger the *apoptosis* response leading to cell death; the other is to employ a special polymerase able to synthesize past the lesion. Most organisms appear to have this ability, which is often mutagenic since it is mediated by *error-prone polymerases. Double-strand breaks (DSBs)* are not uncommon and are repairable, while *DNA strand cross-linking* by acetaldehyde, a by-product of alcohol metabolism, can also be reversed.

MYCO-: **mukes** (Gk.), usually relating to fungi (but see mycobacteria and mycoplasma below), hence *mycotoxin*, a toxin produced by a fungus; *Saccharomyces cerevisiae*: literally 'sugar fungus for making bread', better known as **YEAST** and as a model organism in eukaryotic molecular biology.

MYCOBACTERIA: a genus of bacteria responsible for diseases in mammals including *leprosy* and *tuberculosis* in humans and *Johne's disease*, also known as *paratuberculosis*, in sheep and cattle. The **myco**- prefix is on account of the way these bacteria grow like moulds on the surface when in liquid culture.

MYCELIUM: vegetative, underground growth of fungi.

MYCOPLASMA: **mukes** + **plasma** from **plassein** 'to shape'. A large bacterial genus, many species of which are *obligate parasites* because of gene loss. *Mycoplasma* infect both animals and plants. The misleading name goes back to early observations that were attributed to a fungus. At one time, *Mycoplasma* were known as *pleuropneumonia-like organisms* (PPLOs), stemming from the disease affecting cattle, now known to be caused by *M. mycoides*. *Atypical pneumonia* in humans is caused by *M. pneumonia*.

Because of their small size, less than 1 μm, they are difficult to detect by conventional microscopy and pass through filters that retain other bacteria. Because they have no cell wall, antibiotics such as penicillin that target cell wall synthesis are not effective. Partly for these reasons, mycoplasma became established in many *cell lines* used in research laboratories, a problem that was slow to be appreciated. Detection methods based on DNA and *enzyme immunoassays* are now available, together with antibiotic reagents able to eliminate mycoplasma from cell cultures.

Mycoplasma species have attracted the attention of *synthetic biologists* who are interested in constructing organisms specifically designed to manufacture useful products. One reason for this interest is their relatively small genomes; another is the absence of a cell wall which facilitates DNA uptake. Venter's group succeeded in assembling a completely synthetic *M. mycoides* genome, 1.08 Mb in length, and then transplanting it into the related *M. capricolum*. Subsequent growth in the presence of tetracycline selected for cells containing only the synthetic genome. This work 'may facilitate construction of useful microorganisms with the potential to solve pressing societal problems in energy production, environmental stewardship, and medicine'.[75]

MYCOMYSTICISM: mystical experiences induced by consumption of psilocybin obtained from the *psychotropic* 'magic mushroom' *Psilocybe cubensis.*

MYCORRHIZA: mukes + rhiza 'root', as in rhizome. Fungi that grow in association with the root of a plant. Usually this symbiotic relationship is mutually beneficial, in which case the fungus is said to be *mycotrophic.*

MYELIN: muelos (Gk.) 'marrow': electrically insulating material that forms a sheath around the axons of neurons in the brain and spinal cord. Myelinated axons are white, hence the white matter of the brain. The myelin sheath consists of the *plasma membranes* of cells known as *oligodendrocytes* which has been wrapped multiple times around the axon. Remyelination occurs throughout adult life and involves the generation of new myelinating oligodendrocytes. This area is of research interest because demyelination is characteristic of *multiple sclerosis*. The cause of this disease is unknown, but it is generally considered that it involves an autoimmune attack by the immune system directed against myelin with possible contributions of environmental and genetic factors.

POLIOMYELITIS: polios 'grey' + **muelos**: infectious viral disease affecting the central nervous system. Polio is now *nearly* extinct because of worldwide vaccination efforts.

See also CHRONIC MYELOID LEUKAEMIA under **CANCER**.

NAME: onuma (Gk.). Since everything has a name, this root has given rise to numerous words in English:

Ananym: **ana-** 'back' + **onuma**: a word or words which when read backwards gives another word, like the racehorse called Red Rum; also Llareggub, the name of the Welsh village in *Under Milkwood* by Dylan Thomas.

Acronym: **acro-** 'tip' or 'extremity': a group of letters consisting of the initial letters of words e.g. DNA, SCUBA; see also chapter 3.

Anonymous: **a-** 'not', so literally 'no name', or at any rate, name not supplied and often abbreviated to 'anon.' The corresponding noun is *anonymity* as well as the recently coined verb *anonymize*. Important privacy issues are raised by the advent of large data sets containing medical, financial, and other information together with the computing power to mine them. A study of credit card data shows it takes only a tiny amount of personal information to deanonymize people. We have been warned.

Antonym: **anti-** 'opposite': a word that has the opposite meaning to that of another, e.g. *good* is the antonym of *bad*.

Eponym and the more familiar *eponymous*: giving one's name to something, e.g. Murphy and his eponymous law. Units such as the Dalton, the watt, and the joule are eponyms.

Metonymy: literally 'change of name': a word whose meaning alters, e.g. *apparatus*, where the original meaning meant preparation for a procedure and which in the context of experiments came to refer to the hardware to be used.

Patronymic: name derived from that of a father or ancestor, e.g. a family name as in Johnson.

Pseudonym and *pseudonymous*: a false or fictitious name often used by authors and contributors to social media.

Synonyms: different words that have the same meaning—accordingly, different codons that encode the same amino acid are said to be synonymous.

Tautonym: **tauto-** 'same': a binomial name in which the same word is used for both genus and species, e.g. *Rattus rattus*, the black rat.

Toponym: **topo-** 'place' is a place name, usually based on a particular topographical feature, e.g. Broken Hill.

Then there was the ill-starred attempt to teach a chimpanzee to talk, where the subject was named Nym Chimpsky, a nod to the linguist Noam Chomsky.

The Latin for name is **nomen, nomina**, which provides the root for *nomenclature*, which incorporates **calare, -avi, -atum**, 'to call or proclaim'. The verb corresponding to **nomen** is **nominare, -avi, -atum**, from which we get *nominate* and *denomination*. Also from **nomen** is *ignominy*, meaning public shame. Similarly, *binomial* as in *binomial terminology* for naming species, e.g. *Homo sapiens*, derives from the Latin **binominis** 'having two names'. Contrary to appearance, *phenomenon* is pure Greek.

NATIVE: nasci (L.) 'to be born' and **natus** 'born'. From which are derived *prenatal, nascent, innate, cognate, nature, natural, denatured*. Denatured proteins are proteins that have been unfolded, e.g. by heating or exposure to detergents or extremes of pH, with a consequent loss of their biological activity, from the Latin **denasci** 'to perish or die'. Proteins in good health are referred to as *native*. *Nascent* describes macromolecules still being synthesized. There is also a class of proteins known as *natively unfolded proteins*, which contain substantial regions of intrinsically disordered sequence. Such proteins are often involved in signalling and may undergo structural transformations when they interact with a ligand. *Cognate*: related or connected, as for an amino acid and its cognate tRNA. **Cognatus** in Latin meant 'blood relation', i.e. born together from **co-gnatus, gnatus** being an irregular form of **natus**. In Greek **physis, -ios** (Gk.) meant 'nature or condition' and it is the root for *physics, physician,* and *physiology*; **physein** *v.* 'to grow out of, to appear by itself'.

NOCEBO from **nocere** (L.) 'to harm or injure'. The *nocebo effect* is the opposite of the *placebo effect*. The latter is a well-known effect in drug trials where patients who receive a placebo, i.e. a control that contains no drug, report responses expected of the drug being tested. The nocebo effect occurs when patients have been warned of possible side effects. If the doctor says, 'This may cause you to feel off-colour for a day or two', then the patient may well report that indeed it did. This complicates drug trials where, in order to obtain informed consent, patients need to be warned of possible side effects. See the entry for HOMEOPATHY under **PATHO-**. NOCICEPTOR: a *sensory receptor* that responds to painful stimuli—see under **RECEPTOR** heading.

NUCLEIC ACIDS and GENE EXPRESSION. The term *nucleic* obviously derives from the early identification of DNA within the *nucleus* of cells; **nucleus** was a term used by the Romans, meaning the centre of an object and derived in turn from **nux, nucis,** meaning nut or kernel.

DEOXYRIBONUCLEIC ACID (DNA) is a *polynucleotide*, which is to say it is a long polymer in which the repeat unit is the *nucleotide* consisting of three components: a nitrogen-containing base, the sugar 2-deoxyribose, and a phosphate residue. The four different bases that occur in DNA are adenine and *guanine*, both derivatives of *purine*, as well as two *pyrimidines*, *thymine* and *cytosine*, abbreviated to A, G, T, and C. As is usually the case, these names derive from materials from which they were first isolated: adenine from the **aden** (Gk.) 'gland', guanine from the seabird excrement guano, thymine from **thumos** (Gk.) 'thymus gland', and cytosine from **kutos** (Gk.) 'cell'. An alternative name for thymine is 5-methyl uracil; uracil, along with A, G, and C occurs in *ribonucleic acid* (RNA). Uracil (U) was named for urea and acrylic acid, the starting materials for its original laboratory synthesis. Apart from uracil, the other difference between DNA and RNA is the presence of the sugar *ribose* in place of deoxyribose which has the effect of making RNA less stable than DNA, particularly at alkaline pH. A base plus a sugar is referred to as a *nucleoside*, which can, of course, be differentiated into ribonucleosides and deoxyribonucleosides depending on their sugar components. These are named as follows: *adenosine, guanosine, thymidine, cytidine,* and *uridine* with the prefix deoxy- as appropriate. Nucleotides phosphorylated to varying degrees are named as derivatives of the nucleosides, hence deoxyadenosine monophosphate (dAMP), the diphosphate is dADP and the triphosphate, dATP, or when the sugar is ribose, ATP, the familiar energy currency molecule. For structures of nucleotides, go to https://en.wikipedia.org/wiki/nucleotide.

THE DNA HELIX consists of two *polynucleotide* strands wound around each other to form a *double helix*—the word **helix** meant *spiral* in ancient

Greek. Each strand consists of a *sugar-phosphate* backbone with the four planar bases A, G, T, and C pointing into the centre of the helix where they form specific *base pairs*, G with C and A with T. *Hydrogen bonds* stabilize the base pairs, helping to hold the two strands together while the *planar* base pairs stack on top of each other, providing additional stabilizing *hydrophobic interactions*. The strands of DNA have *polarity*; the ends of each strand are distinguished as either *5'* or *3'*, indicating the deoxyribose positions that are not part of the sugar-phosphate backbone repeat, with the two strands of the double helix having opposite polarity. Because the atoms in the bases are numbered 1, 2, 3, etc., those of the sugars are numbered 1', 2', 3', etc. The bases are attached to the 1' carbon of the sugar by *N-glycosidic* bonds. The mono-, di-, and triphosphates are attached at the 5' position of the sugar.

The DNA helix is referred to as a *right-handed helix*, the convention being that when looking along the helix, it rotates clockwise as it recedes. Another feature of the DNA helix is that it contains alternating *major* and *minor grooves*, the former providing the preferred space for proteins that interact with the helix. The distance along the helix per base pair (bp) is 3.4 Å (Å stands for Ångstrom unit, 10^{-10} metre); the helix repeats every 10.4 bps, giving a *pitch* of 35.4 Å. Given the haploid content of human DNA in round numbers is 3×10^9 (3 billion) bps, then it follows that this corresponds to a total length of about 1 metre of DNA or 2 metres per *diploid* cell. Extending this calculation to the total length of DNA per human individual, by multiplying by the total number of cells, gives an answer in units of billions of kilometres. It helps to digest this fact by remembering that the DNA fibre is exceedingly thin, ~20 Å. (The total number of cells in a human is estimated at 30×10^{12}.)

DNA REPLICATION: from **replicare** (L.) 'to repeat or double up', derived from the verb **plicare** 'to fold, to coil'. The double-helical structure of DNA was proposed in a short 1953 paper by Watson and Crick. Towards the end of that paper was the sentence 'It has not escaped our notice that the specific pairing we have postulated immediately suggests a possible copying mechanism for the genetic material.' This 'copying mechanism' has been worked out in detail for bacteria and is essentially the same in higher organisms, although details there are still the subject of investigation. Replication in higher organisms, in terms of the rate of incorporation of nucleotides at each origin, is slower than in bacteria, reflecting the need to unwind DNA from *nucleosomes*. Eukaryotic chromosomes consist of single, very long, *linear* double-stranded DNA molecules. The largest of the twenty-three human chromosomes contains about 250 million bps, while *E. coli*, a fairly typical bacterium, has a single *circular* chromosome of approximately 4 million bps. Unlike bacteria, which have a single *origin of replication*, each eukaryotic chromosome has many origins of replication.

DNA COMPACTION. The need to accommodate long DNA molecules within a cell with a diameter measuring a few microns has led to estimates of the degree of compaction of the order of two millionfold. The initial stage of compaction consists of the winding of DNA around a core of *histone* molecules, two copies each of four different histones: H2a, H2b, H3, and H4. The resulting structure is termed a *nucleosome* and consists of 147 bps of DNA looped around the histone core 1.65 times. Histones are basic proteins, i.e. they are positively charged on account of their high contents of *lysine* and *arginine* and they form *salt linkages* with the negatively charged phosphate residues of DNA. Nucleosomes are joined by *linker DNA* so that the total DNA per nucleosome is about 200 bps. The linker DNA provides flexibility and the next stage of compaction involves condensation of nucleosomes into a 30 nm *chromatin* fibre, thought to represent the form of most DNA present during *interphase*. This is the stage of the life cycle of the cell at which *gene expression* occurs via *transcription* of DNA by RNA polymerase and when *DNA replication* occurs. *Chromatin* is defined as the complex of DNA, protein, and some RNA which occurs in the nucleus of eukaryotic cells. Chromatin that is decondensed is referred to as *euchromatin*. Micrographs show loops of *nucleosomal DNA*, corresponding to 50–100 kbp of DNA (1 kbp = 1,000 base pairs) protruding from a *central protein scaffold*. These have been interpreted as corresponding to a *modular* arrangement of DNA within the nucleus. About 10 per cent of chromatin remains more compact during interphase. This is referred to as *heterochromatin*, and it includes sequences around the *centromeres*, which consist of highly repeated DNA sequences. DNA replication of chromosomes during interphase results in formation of two daughter *chromatids*, which remain attached to each other at their *centromeres* until, during mitosis, the centromeres become attached to the *microtubules* of the *mitotic spindle* prior to *cytokinesis*. Clearly, as chromosomes become visible in the light microscope prior to mitosis, still further compaction must be involved. See also **MITOSIS** and the **CELL CYCLE**.

TELOMERES and TELOMERASE: **telos** 'end' + **mere** 'body'. These specialized structures at the ends of linear eukaryotic chromosomes are characterized by the presence of many copies of a short *repeated sequence*. In vertebrates, the repeat consists of the sequence TTAGGG, and in humans, the total length of the telomere repeat amounts to several kilobases. This repeat, together with proteins that bind to it, can be regarded as forming a kind of cap on the end of the chromosome which prevents its degradation by *exonucleases* and protects it from recognition by the *DNA repair machinery* which recognizes *double-strand breaks* and which otherwise could result in chromosome translocation. Telomeres are synthesized by the enzyme *telomerase*. In addition to protein subunits, it also contains an RNA *molecule* about 450 nucleotides long that contains near its 5' end the *reverse complement* of the TTAGGG repeat, i.e. CCCTAA. Telomerase, it turns out, is a *reverse transcriptase*, i.e. an enzyme that makes a DNA copy of an RNA sequence, and in this case, it is the RNA component of the enzyme that serves as *template* for the synthesis of successive copies of the TTAGGG repeat.

RIBONUCLEIC ACID and TRANSCRIPTION: **transcribere** 'to transcribe' from **trans-** 'across' + **scribere, scripsi, scriptum** 'to write'. Transcription consists of the synthesis by RNA polymerase of a strand of RNA using one of the strands of double-stranded DNA as *template*, made available by localized transient unwinding of the double helix. Initiation of transcription involves recognition by RNA polymerase of a DNA sequence known as the *promoter*, located in the vicinity of the start-point of transcription. The substrates are the four ribonucleoside 5' triphosphates (NTPs), and RNA synthesis occurs in the 5' to 3' direction. The strand elongation mechanism is similar to that for DNA synthesis.

RNA polymerases (RNAPs) that transcribe from a DNA template are also described as *DNA-dependent RNA polymerases* to distinguish them from enzymes that replicate the RNA genomes of RNA viruses, the *RNA-dependent RNA polymerases*. In bacteria, a single RNAP suffices for transcription of the major RNAs, while in eukaryotes, a division of labour occurs: RNAP I (Pol I) synthesizes ribosomal RNA, RNAP II

(Pol II) transcribes all mRNAs, and RNAP III (Poll III) the tRNAs and other small RNAs. Genes transcribed by Pol III have the distinctive property of containing internal promoters. This has had a profound effect on the composition of the human genome because of the activity of the enzyme *reverse transcriptase*, which is able to synthesize DNA from an RNA template—see entry for RETROTRANSPOSONS under **GENOMES**. Initiation of mRNA synthesis by Pol II in eukaryotic cells is complex, in line with the complexity of the genome and challenges posed by the tissue specificity of transcription. Eukaryotic Pol II consists of ten to twelve subunits. Assembly of the initiation complex depends on recognition by *transcription factors* of upstream DNA sequences. In many cases, efficient transcription also depends on sequences termed *enhancers*, which can be located thousands of base pairs on either side of the transcription start site. The LP allele is an example of this—see entry for DIET and EVOLUTION under **MILK**. Enhancers are short DNA sequences that bind transcription factors, i.e. proteins that regulate transcription. Enhancers are *cis-acting*, which is to say they are located on the same chromosome as the gene whose expression they regulate. Their ability to exert their effects depends on looping of chromosomal DNA to bring the enhancer adjacent to the gene promoter, the binding site for RNA polymerase and transcription factors. Splicing out of introns from the mRNA *primary transcript* occurs in the nucleus, carried out by a ribonucleoprotein complex referred to as the *spliceosome*. See also the HUMAN PROTEOME entry under **PROTEINS**. The mRNA is then exported to the cytoplasm, where protein synthesis occurs.

EPIGENETICS: **epi**- 'in addition' + genetics. Genetics has to do with the sequence of DNA nucleotides, the genotype, and how alterations in genotype affect phenotype. Epigenetics refers to variables in addition to nucleotide sequence which influence transcription: one has to do with DNA modification, mainly the methylation of cytosine residues at the 5 position; the other consists of modifications such as methylation and acetylation of particular amino acid residues of the histone proteins around which DNA is wound in nucleosomes. Taken together, these two types of modification are referred to as the *epigenome*. Unlike the genome, the epigenome is *tissue-specific*.

Methylation of cytosine at the 5 position to form 5mC is the predominant modification of DNA in vertebrates. In mammals, 5mC occurs mainly in CpG dinucleotides, and 70–80 per cent of CpGs are methylated on both strands. Both *de novo* and maintenance methylating enzymes occur. The latter act on newly replicated DNA and recognize the 5mC on the template strand in order to methylate the C on the newly synthesized strand. As a result, DNA methylation can be heritable. Traditionally, the view has been that the role of methylation is to silence transcription. Application of genome-wide methods for detecting DNA methylation have shown that this view is oversimplified and that both methylation and demethylation are themselves subject to regulation depending on the tissue and stage in the life cycle of a cell.

TRANSLATION or protein synthesis is the second stage, after transcription, of gene expression. The word *translation* normally refers to the rendering of one language into another, and this is also the case here because a sequence of nucleotides in mRNA is translated into a sequence of amino acids in a protein. *Translation*, the word itself, derives from the past participle of the irregular verb **transferre** (L.); see chapter 2. Protein synthesis occurs within a ternary complex involving the three major categories of RNA: *ribosomal RNA* (rRNA), amounting to 50 per cent of the ribonucleoprotein particle the ribosome, *messenger RNA* (mRNA) containing in its nucleotide sequence the information for protein amino acid sequence, and *transfer RNAs* (tRNAs), which carry activated amino acids to the ribosome. Each triplet of nucleotides, or *codon*, in mRNA is read by a complementary sequence in tRNA, the *anticodon*, with each tRNA and hence each anticodon being specific for one of the twenty amino acids. There are two steps that ensure the accuracy of protein synthesis: one is the codon-anticodon interaction that relies on specific pairing, on the ribosome, between triplets of bases in tRNA and mRNA; the second relies on a set of enzymes known as *aminoacyl-tRNA synthetases*, specific for each amino acid, that catalyse reactions between each amino acid and its cognate tRNA.

THE GENETIC CODE TABLE (see below) summarizes the rules for translating an mRNA sequence into a protein sequence. The genetic code is *triplet*, i.e. each of the *codons* that specify an amino acid and each of the three *stop* codons consists of three nucleotides. All sixty-four (4×4×4) possible codons are used. The code is *non-overlapping*, and there is no punctuation, i.e. each triplet of bases or codon is followed immediately by the next triplet. The three *stop* codons indicate 'end of message' and the remaining sixty-one codons are distributed between the twenty amino acids. The code is therefore said to be *degenerate*—meaning most amino acids are specified by more than one codon. Two or more codons specifying the same amino acid are said to be *synonymous*. The single codon for methionine, AUG, is also the *initiation codon*. Most proteins are synthesized beginning with methionine, which is later removed. AUG also codes for internal, i.e. downstream methionines. Attempts to find structural relationships between amino acids and their codons or anticodons have not been successful, with the result that the genetic code is regarded as a frozen accident.

TABLE 9. THE GENETIC CODE

UUU	Phe	UCU	Ser	UAU	Tyr	UGU	Cys
UUC	Phe	UCC	Ser	UAC	Tyr	UGC	Cys
UUA	Leu	UCA	Ser	UAA	Stop	UGA	Stop
UUG	Leu	UCG	Ser	UAG	Stop	UGG	Trp
CUU	Leu	CCU	Pro	CAU	His	CGU	Arg
CUC	Leu	CCC	Pro	CAC	His	CGC	Arg
CUA	Leu	CCA	Pro	CAA	Gln	CGA	Arg
CUG	Leu	CCG	Pro	CAG	Gln	CGG	Arg
AUU	Ile	ACU	Thr	AAU	Asn	AGU	Ser
AUC	Ile	ACC	Thr	AAC	Asn	AGC	Ser
AUA	Ile	ACA	Thr	AAA	Lys	AGA	Arg
AUG	Met	ACG	Thr	AAG	Lys	AGG	Arg
GUU	Val	GCU	Ala	GAU	Asp	GGU	Gly
GUC	Val	GCC	Ala	GAC	Asp	GGC	Gly
GUA	Val	GCA	Ala	GAA	Glu	GGA	Gly
GUG	Val	GCG	Ala	GAG	Glu	GGG	Gly

NUMBERS and UNITS

NUMBER: **numerus** (L.), **arithmos** (Gk.). Numbers are classified as either *cardinal* numbers, such as 1, 2, 3, etc. or *ordinal* numbers, 1st, 2nd, 3rd, etc., both derived from Latin **cardinalis** and **ordinalis**, meaning 'chief or fundamental' and 'order' respectively. As indicated below, scientific English has dipped into both Latin and Greek cardinal and ordinal numbers as a source of numerical prefixes.

Numerical prefixes derived from Greek:

ONE: **monos** (Gk.) 'one, single, alone' as in carbon monoxide, monoamine oxidase, monoclonal antibody, monolayer, monomer, monotreme.

TWO: **di** (Gk.) 'two, double' as in diploid, carbon dioxide, disulphide bond, diatomic, dimer, adenosine diphosphate, dipeptide.

THREE: **treis** (Gk.), also **tres** (L.) 'three', to have three of something as in *tri*angle, adenosine *tri*phosphate (ATP). When it comes to chemicals containing three groups, these are named as either *tri-* or *tris-* depending on whether or not the groups are identically substituted. For the phosphates in ATP, the tri- indicates that these phosphates are not identically substituted. ATP can be represented as adenine-ribose-P-P-P. The chemical environment of each of the three phosphates is clearly different. The second messenger molecule IP_3 (inositol containing phosphate residues esterified at the 1, 4, and 5 positions) is referred to as inositol-1,4,5-*tris*phosphate. Another example where **tris** is used is the pH buffer called Tris, which is *tris*(hydroxymethyl)aminomethane, pK_a = 8.07. Formula: $(HOCH_2)_3CNH_2$. Here the three hydroxymethyl groups, attached to the same carbon atom are equivalent. See below the entry for BIS-, where a similar rule applies.

TRISOMY: **tri** + **-some** from **soma** 'body': presence of an extra copy of a particular chromosome, usually resulting in developmental abnormalities. An example is Down's syndrome, caused by an extra copy of chromosome 21.

FOUR: **tetra**- as in tetrad, tetrapod, tetrahedron, tetravalent, tetramer, tetracycline, tetrahydrofolic acid, tetraploid, tetrose.

FIVE: **penta**- as in pentavalent, pentagon, pentamer, pentadactyl, pentane, pentathlon, pentapeptide. SIX: **hexa**- as in hexose, hexane, hexagon, hexaploid, and hexavalent, which refers to elements able to exist in the +6 oxidation state, such as sulphur as sulphates and sulphur hexafluoride, SF_6, also chromium as chromates, CrO_4^{2-}.

SEVEN: **hepta**- as in heptoses, heptad repeats in fibrous proteins, heptahydrate.

EIGHT: **octo**- as in octopus, octagon, octane, octanoic acid. **Octo**-indicates eight in both Latin and Greek.

NINE: **ennea**- as in ennead, meaning a group of nine people or things, cf. dyad, tetrad, heptad.

TEN: **deca**- (both Gk. and L.), as in decade, decagon, decanoic acid.

TWELVE: **dodeca**- as in dodecahedron, sodium dodecyl sulphate; Latin for twelve is **duodeni,** from which *duodenum* derives, it being the first portion of the small intestine immediately below the stomach, the length of which mediaeval anatomists determined to be twelve finger widths.

TWENTY: **eicosa**- as in eicosanoids—see under **FATTY ACIDS.**

Numerical adjectives

FIRST: **protos** (Gk.) 'first' or 'earliest form' as in prototype, protozoa, and proton, the subatomic particle present in all atomic nuclei, also known as the hydrogen ion H^+.

SECOND: *deuterium* from **deuteros** (Gk.) 'second', the second isotope of hydrogen containing one neutron. Strictly speaking, the standard atom of hydrogen with a single proton and no neutrons should be called *protium* from Greek **protos** 'first'. In the early nineteenth century, it was suggested that metal oxides with varying proportions of oxygen be named protoxides, deuteroxides, etc., with the compound having the most oxygen being the peroxide, a convention that lingers on with names such as hydrogen peroxide and the permanganate, persulphate, and periodate ions.

PROTOSTOMES and DEUTEROSTOMES: **stoma, stomat-** 'mouth'. Animals having *bilateral symmetry*, the *Bilateria*, fall into two groups—the protostomes and the deuterostomes. These are distinguished depending on whether, during their embryonic development, the first opening becomes the mouth or the anus. In protostomes, the first opening becomes the mouth, while for deuterostomes, which include the *chordates*, it is the second opening.

THIRD: *tritium*: **tritos** (Gk.) 'third'. Radioactive isotope of hydrogen, symbol ^3H. Half-life: 12.3 years. The tritium nucleus contains one proton and two neutrons.

Numerical prefixes derived from Latin

ONE: **unus** (L.): 'one' as in unicellular, unimolecular, unit cell, univalent. Also, **unicus**: 'one and only one, one and no more, unique'. Ordinal: **primus**.

TWO: **bi-** (L.): 'two': binocular, biped, bisect, bivalve mollusc, biennial, bilateral symmetry, bisubstrate reaction. Also **binarius**: 'that contains or consists of two', hence *binary*. There is an inconsistency in chemical nomenclature that goes back a long way, namely *bi*carbonate for HCO_3^-. Ordinal: **secundus** as in secondary education.

BIS-: **bis** (L.): 'twice' (**bis** and 'twice' are *numerical adverbs*) as in fructose 1,6-*bis*phosphate, not diphosphate. The rule is that **bis-** indicates the presence of two identical groups, each identically substituted. On the other hand, ADP, adenosine 5'-*di*phosphate, is so named because the two phosphates are not identically substituted. However, it is 1,3-*bis*phosphoglycerate, the glycolysis intermediate, and sedoheptulose 1,7-*bis*phosphate, an intermediate in the Calvin cycle and the pentose phosphate pathway. I am told that enthusiastic French audiences shout 'bis!' rather than the French word 'encore!' which is what English audiences shout.

THREE: **tri-** (L.), same as the prefix derived from Greek—see above. Ordinal: **tertius** as in tertiary education.

FOUR: **quattuor** (L.). Ordinal: **quartus** as in quart, quarter.

FIVE: **quinque**: (L.) as in quinquennial: of or for five years. Ordinal: **quintus** as in quintet, quintuplicate.

SIX: **sex** (L.). Ordinal: **sextus**.

Latin numbers seven to ten or **septem, octo, novem,** and **decem** (ordinals: **sextus, septimus, octavus,** and **decimus**) tend not to occur in the scientific lexicon but are familiar as names for the months September, October, November, December. The Roman calendar, like the modern calendar, had twelve months, but the first month was March.

PROTEIN STRUCTURE is described as above:

Primary: **primus** (L.): the sequence of amino acids

Secondary: **secundus** (L.): regular structures such as the α- helix and β-pleated sheet

Tertiary: **tertius** (L.): the three-dimensional folded structure

Quaternary: **quartus** (L.): associations yielding homo- and heterodimers, tetramers, etc. See also under PROTEINS.

Numerical adjectives derived from Latin

SINGLE: **singulus** (L.) as in single-stranded DNA; carbon-carbon single bond as in ethane.

DOUBLE: **duplus** (L.). Double-stranded DNA; carbon-carbon double bond as in ethylene.

TRIPLE: **triplus** (L.). Carbon-carbon triple bond as in acetylene. QUADRUPLE: **quadruplus** (L.). Fourfold, as in quadruplets.

As above, but derived from Greek:

SINGLE: **aploos,** as in *haploid*, referring to a single set of unpaired chromosomes as in gametes; also in *haplotype*, the particular combination of alleles present in a specific region of a chromosome.

DOUBLE: **diploos,** as in diploid set of chromosomes in somatic cells.

TRIPLE: **triploos,** as in triploid.

QUADRUPLE: **tetraploos,** as in tetraploid. See also under PLOIDY.

SIMPLEX: **simplus** (L.): a term applied to something composed of a single component; cf. MULTIPLEX, which can refer to an analytical system processing many samples at once and/or measuring multiple properties.

DUPLEX: from **duplex, duplic-**. Most commonly encountered as 'duplex DNA', the sense being something composed of two components, for DNA composed of two strands; also *duplicate*, meaning, as a verb, to make another copy while, as a noun, the duplicate, pronounced with a short a, *is* the other copy.

Other miscellaneous words that belong here include *quarantine*: originally a period of forty days of isolation imposed on ships suspected of carrying disease, from the Italian **quarantina**.

QUARTAN FEVER: a mild form of malaria that recurs every fourth day, from **febris quartana,** from **quartus** 'fourth'. Confusingly, this type of malaria gives rise to a fever every third day, but is described as every fourth day by inclusive reckoning.

HUNDRED: **centum** (L.): *cent, centimetre, century, percentage, centigrade, centenary*; **hekaton** (Gk.) 'hundred', hence *hectare* (100 metres × 100 metres = 10,000 square metres = 1 hectare, the metric unit of land area).

THOUSAND: **khilios** (Gk.), *kilogram* (1,000 grams), *kilometre*; **mille** (L.): *milligram* (one thousandth of a gram), *millimetre, millennium*, and *million*, i.e. a thousand thousands. See also table 10 below.

UNITS

ANGSTROM: unit of length equal to 1×10^{-10} of a metre or 10^{-1} of a nanometre (nm), symbol: Å, and named after the pioneer Swedish physicist Anders Ångström. This is a useful unit for describing biological molecules. For the structure of DNA, base pairs repeat every 3.4 Å along the helix, which is 20 Å in diameter. The complete determination of protein structure by X-ray diffraction or cryo-electron microscopy requires a resolution of about 3Å.

AVOGADRO'S NUMBER: 6.023×10^{23} is the number of units (atoms, molecules) in one mole of any substance where the *mole* is defined as the atomic or molecular weight in grams.

CALORIE: **calor** (L.) 'heat', a unit of energy, defined as the energy required to raise the temperature of 1 gram of water by 1°C. Because the energy required depends slightly on the starting temperature, the calorie as used in biology is defined as the energy required to raise the temperature of 1 gram of water from 14.5°C to 15.5°C. The physical sciences have mainly abandoned the use of calories in favour of the SI unit of energy, the *joule*. One calorie = 4.2 joules approximately. The biological literature still retains calories, with the free energy of hydrolysis of ATP to give ADP and Pi given as -7.3 kcal/mole under standard conditions. (Pi is shorthand for inorganic phosphate.)

MOLE: the molecular weight of a compound expressed in grams.

DALTON: the unofficial but widely used unit for the mass of molecules is the Dalton (Da), where 1 Da is the mass of the ^{12}C isotope of carbon, divided by 12. John Dalton (1766–1844) is regarded as a major figure in the development of modern chemistry through his championing of the atomic theory of matter as well as for many other contributions.

TEMPERATURE SCALES. Lord Kelvin in 1848 introduced the absolute temperature scale in which *absolute zero* or 0°K is the theoretical state in which all molecular motion—translational, rotational, and vibrational— ceases. By international agreement 0°K equals -273.15°C. This figure was arrived at by extrapolating the Gas Law relating temperature, pressure, and volume of an ideal gas and assuming an ideal gas occupies zero volume at 0°K. Where temperature T occurs in thermodynamic equations, it is the Kelvin scale temperature that applies. See also entry for **FREE ENERGY**. OTHER TEMPERATURE SCALES: The Celsius scale is named after Anders Celsius (1701–1744) who invented a temperature scale containing 100 steps or degrees between the temperatures at which ice melted and water boiled, hence also known as the *centigrade* scale (**centum** 'one hundred' + **gradus** 'step'). Celsius originally set the boiling point of water at 0° and the melting point of ice at 100°. It was his compatriot Linnaeus who convinced Celsius to reverse the scale, and it is this which is now used in most countries to express temperatures (symbol: °C). In the other temperature scale, named after Daniel Fahrenheit (1686—1736), ice melts at 32°F and water boils at 212°F. His reasons for setting the 0° mark on his scale are obscure, while the 212° figure apparently enabled him to obtain a figure for the temperature of the human body around 100°F.

Table 10 below lists the base SI units (Système Internationale) and the prefixes used to indicate quantities expressed in terms of these units.

TABLE 10. SI UNITS and their PREFIXES

Base SI Units

SI Unit	Quantity	Symbol
length	metre	m
mass	kilogram	kg
time	second	s
electric current	ampere	A

temperature	kelvin	K
amount of substance	mole	mol
luminous intensity	candela	cd

Prefixes

deci-	d	10^{-1}	deca-	da	10^{1}
centi-	c	10^{-2}	hecto-	h	10^{2}
milli-	m	10^{-3}	kilo-	k	10^{3}
micro-	μ	10^{-6}	mega-	M	10^{6}
nano-	n	10^{-9}	giga-	G	10^{9}
pico-	p	10^{-12}	tera-	T	10^{12}
femto-	f	10^{-15}	peta-	P	10^{15}
atto-	a	10^{-18}	exa-	E	10^{18}

Thus 1 millimetre = 10^{-3} metre = 1 mm. 1 micromole = 10^{-6} mole = 1 μmol, etc.

OESTRUS: oistros (Gk.) 'gadfly'. The names of the female hormones *oestrogen* and *oestradiol* derive from **oistros**. In ancient Greek, various metaphorical meanings were associated with this word, such as a sting that drives to madness or more generally, madness or passion. In current English usage, the word denotes the stage in the *ovarian cycle* prior to *ovulation* where mammalian females of species subject to an oestrous cycle are sexually receptive. Animals with an oestrous cycle reabsorb the *endometrium* if conception does not occur, while those with a menstrual cycle shed the endometrium through menstruation.

The connection with the gadfly goes back to Greek mythology, where the story, or one of them, goes that Zeus, the top god, was having it on with the maidservant Io, but Hera, Zeus's wife, was not happy about this, so Zeus, in order to protect Io from Hera, changed Io into a heifer. Hera did not fall for this trick and proceeded to infest Io with the gadfly. Zeus responded by transferring Io across a waterway, and this is the origin of the name for the Bosporus, the narrow waterway joining the Black Sea and the Sea of Marmara: **bos** 'cow' and **poros** 'ford or ferry'. Thus, Bosporus

= ox ford, which happens to also be the origin of the name of a university in the UK although another university, Cambridge, of similar antiquity, evidently managed to have a bridge over the Cam, possibly reflecting superior technical know-how.

-**OME**: Greek suffix meaning 'having the nature of'. It was in 1920 that a botanist, Hans Winkler, suggested the term *genome* to describe a set of chromosomes.[45] In contemporary usage, *genome* refers to the complete haploid DNA sequence of an organism. *Annotation* of a genome sequence yields an estimate of the total number of protein coding genes, in other words the *proteome*, as well as non-coding sequences that regulate gene expression. Because many proteins interact with each other in the living cell, the total interactions in a given cell type constitute the *interactome*. The total RNA sequences, coding and non-coding, expressed in a particular type of cell under a particular set of conditions, is the *transcriptome*. The epigenetic modification of DNA and histones, when analysed across an entire genome, is the *epigenome*. The set of low-molecular-weight metabolic compounds is the *metabolome*. Neurons that connect with each other constitute the *connectome*. The range of bacterial species found in the human stomach is the *gut microbiome*, while the set of several hundred protein kinases has been dubbed the *kinome*. The set of all known viruses is the global *virome*. As with the genome, most of these other '-ome'-type fields depend on sensitive, high-throughput techniques that generate large amounts of data. A certain amount of scepticism has greeted the rapid generation of large numbers of biological terms with *-ome* on the end: 'By virtue of that suffix, you are saying that you are part of a brand new exciting science'.[76]

-**OMICS**: The general area of genome research became known as *genomics*, so it follows that there are the fields of *epigenomics*, *proteomics*, *metabolomics*, etc. An -omics area that will keep biologists employed for the foreseeable is *phenomics*. Here the aim is to correlate the properties of an organism, in other words its *phenotype*, with its *genotype* or genome. As well as for humans, phenomics projects are underway for most of biologists' favourite model organisms, including yeast, mice, zebra fish, and the plant *Arabidopsis thaliana*. The approach, at least initially, is to inactivate genes one at a time and to look for effects on physical form, metabolism, behaviour and responses to stress. See also entry for CRISPR under **GENOMICS**.

OPSINS: opsis (Gk.) 'sight', hence *optic* as in *optic nerve, optical* and *optometrist, biopsy*. Opsins mediate the initial events in the *retina* that generate *nerve impulses* that result in vision. But opsins also occur in organisms that are regarded as representatives of the earliest *metazoans*, and it appears that they are ancient *sensory proteins* that serve functions in addition to their ability to sense light. All opsins that function in *photoreceptor cells*, i.e. those which sense light, bind *11-cis-retinal* covalently via a *Schiff base* between the *aldehyde* group of retinal and the side-chain *amino group* of a conserved *lysine* in the opsin receptor (R_1-CHO + R_2-NH$_2$ → R_1-CH=N-R_2). Opsins are members of the 7TM G protein-coupled receptor (GPCR) superfamily (see under **RECEPTOR** heading). Rather than the binding of a *ligand*, as with other GPCRs, it is *isomerization* of 11-cis-retinal to *all-trans retinal* by light that leads to a *conformational change* in the receptor and activation of a *second messenger cascade* and eventual generation of a nerve impulse.

RHODOPSIN: **rhodon** 'rose', originally called *visual purple*, is the light-sensitive pigment present in the *rod cells* of the vertebrate retina and is responsible for *monochromatic* vision under conditions of low light intensity. In humans and other animals with colour vision, the *cone cells* of the retina contain three different opsins. The retinal *prosthetic groups* in these three opsins absorb light at different wavelengths because they are located in slightly different environments due to variation of the amino acid sequences of the opsins. These three opsins are referred to as red, green, and blue according to the wavelengths of their absorption maxima. This is the basis for colour vision. Cone opsins are sometimes referred to as *photoopsins* to distinguish them from rhodopsins, while opsins that bind retinal are also referred to as *retinylidene proteins*.

Rhodopsins also occur in bacteria, where they serve as light-gated ion channels, i.e. they are activated by light. *Bacteriorhodopsin* functions as a *proton pump, halorhodopsin* as a *chloride pump*. These opsins are not G protein-coupled. There are nine opsins encoded in the human genome; one, *melanopsin*, is thought to play a role in *circadian rhythm*, others, referred to as *encephalopsins*, are expressed in the brain and their function is unknown. The fruit fly *Drosophila melanogaster* has seven opsins, some involved in vision, another is expressed in *mechanosensory cells* of the fly ear and is required for hearing, while yet another fly opsin is involved in *temperature sensing.* That opsins have a long evolutionary history is shown by examples mediating light sensing present in sea urchins and jellyfish.[77]

ORGAN, ORGANIZE, ORGANELLE, ORGANOIDS, and ORGANISM

Organ and related words have a long history. Aristotle (384–322 BCE) used *Organon* as the title of a series of logical treatises in which **organon** was the hypothetical mechanism or instrument of all reasoning. In the hands of the Romans, a more practical people, **organon** became **organum**, referring to implements and engines, usually of the military variety. A secondary meaning that has survived from around this time refers to instruments for making music. Fast-forward to 1620, and Francis Bacon (1561–1626) publishes *Novum Organum*, a new *Organon*, described as the canonical exposition of the scientific method. Given that it was in 1628 that William Harvey published 'On the Motion of the Heart and Blood', the time was ripe to think in terms of *organisms* as composed of *organs* and to invent the word *organized* to describe the nature of organs and organisms. (According to the *OED*, the first use of *organized* was in 1598). The nineteenth century saw the rise of *organic chemistry*, the chemistry of carbon compounds, while the twentieth saw the mildly annoying habit of food grown without added fertilizer being referred to as 'organic food'. The invention of the electron microscope revealed structures within cells known by the diminutive as *organelles* such as *mitochondria*, *lysosomes*, and *endoplasmic reticulum*.

ORGANOIDS consist of tissue samples cultured in *microfluidic devices* designed to maintain conditions that allow biopsy samples to continue to grow. The key advance here is that tissues grow in three dimensions under precisely defined culture conditions. One of many applications of this technique is to use multiple tumour-derived organoids from a cancer patient to screen for the effectiveness of *antitumour* drugs for that particular tumour. This then allows optimal tumour-specific *chemotherapy*.[78]

MODEL ORGANISMS. Much progress in science and medicine has relied on experiments with model organisms. Some of these are

Drosophila melanogaster: easy to propagate, with visible banding of salivary gland chromosomes because of multiple rounds of DNA replication without mitosis and cell division; a continuing favourite since the early days of genetics.

Bacteria and various bacteriophages: Jacob and Monod laid the groundwork for understanding the regulation of microbial gene expression using *E. coli* and λ bacteriophage—see Jacob (1995).

Saccharomyces cerevisiae: yeast, a single-cell organism containing a nucleus that has served as a simple model *eukaryote*. Given that about a billion years of evolution separate yeast and humans, one might wonder to what extent yeast serves as a model organism for humans. Kacharoo et al.[79] have addressed this question, and the answer is surprising. They took 414 protein-coding genes that are essential for life in yeast and that have a single corresponding gene in humans, i.e. an *ortholog*, and tested to see if the human gene could replace the yeast gene. The result was that 43 per cent of these human genes were functional in yeast. It was found that seventeen out of nineteen genes of the sterol pathway were replaceable, indicating that interactions between the enzymes of a metabolic pathway constrain genetic divergence.

Caenorhabditis elegans: a tiny roundworm, a *nematode* (**nema** Gk. 'thread') another simple eukaryote, containing just 959 cells and a very simple nervous system or *connectome* consisting in the hermaphrodite of 302 neurons and ~7,000 connections in its neural wiring diagram. The human brain is estimated to contain 86 billion neurons and 100 trillion synapses.[80]

Mouse, *Mus musculus*: model vertebrate and mammal, the experimental animal most frequently used for experiments relating to human medicine.

GENETICALLY MODIFIED ORGANISMS (GMOs) are organisms in which a foreign gene, referred to as a *transgene*, has been incorporated into the genome of a different organism. For animals, the procedure consists of microinjection of DNA into the *pronuclei* of a fertilized egg which is subsequently implanted into the oviduct of a *pseudopregnant* surrogate mother. The result is a *transgenic* animal. Simpler procedures are used to produce genetically modified plants. The aim has often been to improve the nutritional value of food such as the incorporation of *carotene* into rice grains or the production of ω-*3 fatty acids* in pigs.

OSTEO-: **osteon** (Gk.) 'bone', OSTEOBLAST: **osteo**- + **blastos** 'bud or sprout'. Osteoblasts are cells responsible for bone formation. They secrete *osteocalcin* (see below) and *type I collagen*, the main collagen type associated with bone.

OSTEOCALCIN: **osteo**- + calcin = calcium binding. Osteocalcin is secreted by osteoblasts as a precursor protein, which is converted to a peptide containing forty-nine amino acids. It undergoes a *vitamin K–* dependent modification in which three glutamic acid residues are converted to γ-*carboxyglutamic* residues (three-letter code: Gla). This modification results in binding of five calcium ions, and the 3D structure shows that these are located on a surface which is complementary to that of *hydroxyapatite* of bone. The precise role of osteocalcin is not understood, but it is assumed that it has to do with bone mineralization or its maintenance. γ-Carboxyglutamic occurrence is restricted to just three groups of proteins: several proteins in the blood coagulation cascade including *prothrombin*, two proteins, osteocalcin and a matrix protein that are associated with calcified tissue, bone, and teeth, and finally, the vitamin K-dependent enzyme that modifies glutamyl residues, γ-*glutamyl carboxylase*. The effect of an additional carboxyl group at the γ-position is to convert the Glu side chain, which has low affinity for Ca^{++}, into one with high affinity.

OSTEOCLAST: **osteo**- + **klastos** 'broken in pieces'. Osteoclasts are responsible for bone *resorption*. They are large cells containing fifteen to twenty close-packed nuclei.

OSTEOPOROSIS: **osteo**- + **poros** 'passage or pore'. This condition is the result of age-related reduction in bone density, which in turn depends on the balance between bone synthesis and resorption, mediated by osteoblasts and osteoclasts respectively. Both these processes are tightly controlled by hormones so that the loss of oestrogen in women after menopause leads to a decrease in bone density and increased incidence of fractures.

OXYGEN. Priestley is credited with discovering oxygen, although it later became apparent that Scheele had also prepared oxygen three years previously without publishing his work. Lavoisier has also been credited with the discovery, but it is now clear that he knew of the work of one or both of the others. Lavoisier concluded incorrectly that oxygen was the characteristic component of acids and therefore named it: **oxys** (Gk.) 'sharp' + **-gen**: the acid former. Later characterization of *hydrochloric acid*, HCl, showed this was wrong. Hydrochloric acid was known in those days as *muriatic acid*, evidently because it contains chlorine, **muria** being Latin for 'brine', i.e. a concentrated solution of NaCl. The **oxys** root also occurs in *oxymoron*: **oxys** + **moros** 'dull', meaning a seeming contradiction. Examples: *working holiday, military intelligence, rap music, civil engineer, open secret*. Other words containing the **oxys** stem are *paroxysm* (a sudden increase in the severity of a disease or pain or a coughing fit), *oxyntic* cells of the gastric mucosa that secrete hydrochloric acid, the lancelet *amphioxus* 'sharp at both ends'.

OXIDATION. The discovery of the role of oxygen showed that combustion consisted of oxidation, but it was not till the discovery of the *electron* that it was possible to understand oxidation-reduction reactions that do not directly involve oxygen. In Fehling's test for reducing sugars, Cu^{++} is reduced to Cu^+ by addition of an electron at the same time as the *aldehyde* group of the sugar is oxidized to a *carboxyl* group. The stepwise oxidation of ethane can be represented as follows:

$$CH_3CH_3 \rightarrow CH_3CH_2OH \rightarrow CH_3CHO \rightarrow CH_3COOH$$

| Ethane | Ethanol | Acetaldehyde | Acetic acid |

The complete oxidation of glucose *in vivo* consists of multiple steps of intermediary metabolism involving oxidation-reduction reactions, but it is only at the very last step in the *electron transport chain* that electrons reduce oxygen to water. See also under **MITOCHONDRIA**.

OXYGENATION is not the same as oxidation. Haemoglobin is oxygenated in the lungs and deoxygenated in the tissues. Oxygen binds to the iron atom of the haem group of haemoglobin where the iron atom in haemoglobin is in the Fe^{2+} or ferrous state and its oxidation state is not altered by oxygenation or deoxygenation. If the haem iron is oxidized to Fe^{3+}, then haemoglobin can no longer be oxygenated. Normally, blood contains about 2 per cent of haemoglobin in the Fe^{3+} form.

OZONE, O_3, an *allotrope* of oxygen, was first identified in 1840 by Schonbein, who, on account of its smell, thought originally that it might be a new halogen. UV light of wavelength less than 240 nm forms ozone from oxygen. The characteristic smell is often detectable near UV spectrophotometers. It was named for the Greek word **ozo** meaning 'smell', while the element *osmium* was derived from **osme**, also meaning smell, as was *anosmia*, the inability to smell.

PAEDIATRICS: pais, paid- 'child' + **iatros** 'physician': the branch of medicine specializing in the treatment of children. The Greek stem **pais, paid**—meaning 'child or boy' occurs also in paedophile (**paid-** + **philein** 'to love'), denoting someone who engages in criminal sexual conduct with children, i.e. a *paederast*, **paid-** + **erastes** 'lover'. Another word with the **pais** stem is *pedagogue*, which means 'teacher', **paid-** + **agogos** 'leading'. Related words: *demagogue* and *synagogue*. Demagogue: **demos** 'the people' + **agogos**, literally 'leading the people' has traditionally referred to a leader who appeals to prejudice, one liable to lead the people astray; **demos** + **kratia** 'power' is regarded as much safer politically. But sometimes the latter leads to the former. You will notice that following *OED2*, I spell *pedagogue* with an *e* rather than the digraph *ae*, which is or was British usage. The American convention is to spell all such words with an *e* and usage indicates that, with the possible exception of *paediatrics* and *orthopaedics*, which are pronounced with a long *e*, the American spelling will become standard.

There are opportunities for confusion between words based on the **pais** stem, such as pedagogue and words derived from the Latin **pes, ped-** meaning foot. (In *Fawlty Towers*, Basil refers to doctors attending a paediatrics conference as 'foot doctors'). Words derived from the Latin root include *pedestal, pedestrian, pedal*, also *impede*, which comes from **impedire**: 'to shackle the feet', as well as *expedite* and *expedition*. And then there is *podiatry*, the treatment of foot complaints, which comes from the Greek word for foot, **pous, pod-** + **iatros** 'physician'. *Tetrapod* comes from the same Greek stem, while it is *quadruped* in Latin, but if you get along on just two feet, then you are *bipedal*, which derives from the Latin **bipes, biped-**. A matter of contention is the plural of *octopus*. Some people say *octopuses*, others *octopi*. Although it looks like it might be Latin, this *-pus* is Greek, and if you insist on being correct, it has to be *octopodes* (like the *antipodes*) but the best idea is to settle, sheepishly (one sheep, two sheep), for *octopus*. And the same goes for two platypus, (**platus** 'flat'+ **pous**). A Latinized version of the **pous, pod**- stem has come across to English as *podium*, a small platform for a speaker. Meanwhile, single-cell amoeboid organisms move by distending their plasma membranes, involving the assembly of *actin* microfilaments. These cytoplasmic bulges are known as *pseudopodia*, literally 'false feet'. **Pes** and **pous**, as well as OE *foot*, are all related to the Indo-European root **pad**.

PATHO-: **pathos** (Gk.) 'suffering or disease'. The original Greek word could refer to passion or emotion or the passive state, and English words that reflect these meanings include *pathos* itself, *sympathy*, *empathy*, *antipathy*, *apathy*, *pathetic*. Most other words deriving from **pathos** refer to aspects of disease. Usually the stem occurs as a prefix as in *pathogen*: an agent causing disease, or *pathology*: the study of disease. It can also occur as a suffix as in *nephropathy*: kidney disease, also *psychopath*. It is possible to find it used as both prefix and suffix in the same sentence, in this case the title of a journal article: 'Small-fibre neuropathies—advances in diagnosis, pathophysiology, and management'. *Coeliac disease* is sometimes referred to as gluten *enteropathy*; also *myopathy*, *encephalopathy*, *haemoglobinopathy*, and many more.

HOMEOPATHY: **homoios** 'like, resembling' + **patheia** 'suffering'. Homeopathy comes under the heading of alternative medicine, and it was founded in the late eighteenth century by Samuel Hahnemann, a German medical graduate who, with good reason, had become disillusioned with the effectiveness of medicine as practiced at that time. His basic idea was that disease could be cured by administering substances that produced symptoms similar to those of the disease. 'Like cures like' or '**similia similibus curentur**', as he expressed it. Because many of the remedies that conventional medicine was then using involved toxic substances such as mercury, arsenic, and belladonna, Hahnemann began diluting his medicines and eventually decided that the more they were diluted, the better they worked. Modern homeopaths also make this claim, and there is little doubt that their remedies are totally inactive. One concludes that the extent to which homeopathy works is due to the *placebo effect*—**placebo**: (L.) 'I will please'—a phenomenon not unknown in conventional medicine.

Burnside suggests an example of homeopathy in action: 'truth in politics'—the more it is diluted, the better it works.

PEPTIDOGLYCAN. Bacterial cell walls are composed of *peptidoglycan*, which, as the name suggests, consists of short peptides that cross-link polysaccharide chains composed of alternating residues of N-acetyl-D-glucosamine and N-acetyl-muramic acid, joined in β-1,4 linkages. N-acetyl-muramic acid (**murus**, L. wall) is a derivative of N-acetyl-glucosamine, which contains a *lactic acid* residue in an *ether linkage* at the 3 position. Cross-linkage to peptides occurs via an amide bond to the lactic acid carboxyl group. Peptidoglycans contain two separate families of peptides, which are themselves cross-linked by a *transpeptidase*. It is this enzyme which is inhibited by *penicillin*, resulting in a defective cell wall and *cell lysis*. Lysozyme, which cleaves β-1,4 linkages, also targets peptidoglycans. The bacterial cell wall can be regarded as a single huge macromolecule. It provides mechanical strength and rigidity and prevents cell lysis due to high internal osmotic pressure.

PHAGE, -PHAGY: derived from **phagein** (Gk.): 'to eat'. The word *phage* is an abbreviation of *bacteriophage*, or bacterial virus, an appropriate term in view of the ability of bacteriophages to replicate within and then lyse their bacterial hosts. Bacteriophages served as useful model organisms in the early days of nucleic acid research and later as cloning vectors. MACROPHAGY is the term sometimes used to describe the ability of jawed vertebrates, the *gnathostomes* (**gnathos:** 'jaw' + **stoma:** 'mouth') to feed upon organisms larger than those that were preyed upon by their jawless ancestors, the filter feeders.

AUTOPHAGY: **auto** 'self' + **phagein** (Gk.). This is a tightly regulated catabolic process involving degradation of the cell's own components by the lysosomal machinery. It is a mechanism whereby a starving or damaged cell reallocates resources as well as mediating normal turnover of cellular components. Autophagy was originally characterized in yeast cells and later shown to be a fundamental process operating in all eukaryotic cells. In addition, it is now recognized as a key mechanism that can be activated to dispose of invading *pathogens*. Antimicrobial autophagy is also referred to as *xenophagy* (**xenos:** Gk. 'stranger'). Autophagy involves the envelopment of portion of the cytosol in a double membrane-bound vesicle or compartment known as the *autophagosome*, which then delivers its contents to another membrane-bound vesicle, the *lysosome*. Lysosomes maintain a pH around 4.5 and contain up to fifty degradative enzymes, such as nucleases, proteases, lipases, collectively referred to as *acid hydrolases*.

MACROPHAGES are *immune* cells, which is to say they are part of the *innate* immune system and they are found in tissues rather than the circulation. Macrophages originate from *monocytes*, which in turn derive from *haematopoietic* stem cells in the bone marrow. They may be similar in size to their precursor monocytes or they may be *multinucleate*. *Transcriptional profiling*, i.e. mRNA analysis, shows there are many different types of macrophage. Their main function is to maintain *tissue homeostasis* by ingestion of cellular debris but also to engulf invading microorganisms by a process known as *phagocytosis*. Macrophages are able to expand locally in response to infection and are thus an exception to the rule that differentiated cells withdraw from the cell cycle and rely for renewal on stem cells.

PHARMACO-: **pharmakon** (Gk.) 'drug or medicine' and hence 'pharmacist', one who dispenses same. The closely related and presumably earlier Greek word **pharmakos**, according to Liddell and Scott, means a poisoner, sorcerer, or magician. Another meaning is one who is sacrificed as an atonement for others, i.e. a scapegoat, and since worthless fellows were reserved for this fate, **pharmakos** became a general term of reproach or abuse. Interestingly, two-thousand-odd years later, big pharma is often spoken of in similar terms.

PHARMACOGENETICS: the branch of pharmacology concerned with the variable reactions of individuals to drugs as a result of genetic variation, an area of increasing importance. PHARMACOGNOSY: -**gnosis** 'knowledge' the study of drugs derived from plants and other natural sources which, in the era BC (i.e. before chemistry), were the only medicines available. Naturally occurring compounds still provide the starting material for a large proportion of the drugs in current use. PHARMACOLOGY: the study of the biological effects of drugs and their medical applications. PHARMACOPOEIA: -**poiesis** 'making'. An official publication listing current drugs with directions for their preparation and use.

PHARMACOPHORE: **-phoros** 'carrying or bearing'. The particular chemical group responsible for the biological effect of a drug.

PHEROMONE: pherein (Gk.) 'to bear' + **mone** as in hormone. Pheromones were originally defined by Karlson and Luscher[81] to mean a chemical that, when emitted from one animal, exerts a behavioural or physiological response in another animal of the same species. Pheromones thus provide a means of communication based on smell and, obviously, consist of volatile molecules. They can be grouped according to the response they induce, including alarm pheromones, food trail pheromones, mating pheromones, and many others. They can also be classified according to whether the response is immediate or delayed. An example of the former is a pheromone contained in rabbit milk, 2-methylbut-2-enal, $OHC-C(CH_3)=CH-CH3$, which elicits nipple-search behaviour in rabbit pups and is vital for them to locate the nipples during the brief daily period of suckling. Because humans rely heavily on visual and auditory means of communication and, compared to other animals, less on the sense of smell, the question arises whether humans also respond to pheromones. There is some evidence that they *can* but whether in practice they *do* has not been established.

PHORESY: from Greek **phoresis** 'being carried', as in *electrophoresis*, defined in *OED2* as an association between two organisms in which one, e.g. a mite, travels on the body of another, without being a *parasite*. It turns out that Charles Darwin was interested in phoresy, and by a curious coincidence, so was Walter Crick, grandfather of Francis Crick. Matt Ridley, writing shortly after Francis Crick's death, relates how Darwin's final paper, published in *Nature* on April 6, 1882, 'On the dispersal of freshwater bivalves' came about because Walter Crick had written to Darwin to say that he had found a small freshwater cockle, a bivalve mollusc, attached to the leg of a water beetle. Darwin was always interested in how freshwater animals, and molluscs in particular, dispersed by hitch-hiking on other animals. As Ridley notes, Darwin's 1882 paper which appeared just thirteen days before he died was, in effect, a joint publication between Darwin and Crick's grandfather. Walter Crick died in 1903, fifty years before Francis Crick and James Watson published the structure of DNA.[82]

PHOSPHORUS: **phos, phot-** (Gk.): 'light' + **-phorus**: literally 'light bearer', an appropriate name for element number 15, since the white form of phosphorus glows in the dark and is spontaneously flammable when exposed to air. The Latin word **Phosphorus** referred to Venus, the morning star, the 'light bringer'; the word and its meaning they borrowed from the Greeks. In Latin, Venus was known as Lucifer.

Phosphorus occurs naturally only as the phosphate, i.e. a phosphorus atom surrounded by four oxygens. ORTHOPHOSPHATE refers to the ions from phosphoric acid H_3PO_4. At physiological pH, this consists of a mixture of $HPO_4^=$ and $H_2PO_4^-$, the pK_a values of phosphoric acid being 2.2, 7.2, 12.7.[83] Orthophosphate is often abbreviated P_i, the i standing for *inorganic*, to distinguish it from phosphate that is organically bound, such as the phosphate in ATP or glucose-6- phosphate.

Life would not exist without phosphorus. The energy currency of the cell is ATP, adenosine triphosphate. DNA and RNA are polynucleotides, and each nucleotide consists of a base, a sugar, and a phosphate residue. ATP drives muscle contraction, while GTP hydrolysis supplies the energy for the molecular motors kinesin and dynein that are concerned with intracellular transport and determination of cell shape. ATP and GTP have major signalling roles: ATP and to a lesser extent GTP, are substrates for kinases that regulate cell metabolism via both serine/threonine kinases and tyrosine kinases. The GTP/GDP cycle is central to the operation of the several hundred human G protein-coupled receptors (GPCRs)—see entry below for **RECEPTORS**. ATP serves to activate amino acids for protein synthesis by reacting to form aminoacyl adenylates; UTP similarly activates glucose for glycogen synthesis, while CTP plays a similar role in synthesis of *phospho*lipids, a major class of membrane lipids. The average human contains 780 grams of phosphorus, most of it in the form of calcium phosphate in crystals of hydroxyapatite in bone.

PHOTOSYNTHESIS The primary source of nearly all biological energy is photosynthesis, which uses the energy of visible light to decompose water into oxygen, electrons, and protons. In terms of oxidation/reduction, the functions of mitochondria and chloroplasts are the exact opposite of each other. In mitochondria, the reducing power of NADH ultimately reduces oxygen to water, while in the chloroplast, light energy splits water into *oxygen*, *protons*, and energetic electrons able to reduce NADP⁺ to NADPH. Where mitochondria and chloroplasts resemble each other is that both synthesize ATP by a similar mechanism involving formation of a proton gradient. Also, both organelles arose as the result of separate endosymbiotic events. In the case of the *chloroplast*, the invader is believed to have been an ancestor of *cyanobacteria*, also known as *blue-green algae*. Photosynthesis must be very ancient; nevertheless, it appears that life emerged in its absence. The main argument for this is that photosynthesis does not occur among the Archaea, indicating that it began after the divergence of Archaea and the Eubacteria. The ability of existing bacterial groups to use reduced compounds of *sulphur* such as *hydrogen sulphide*, or *hydrogen* itself, as a source of reducing power is consistent with early forms of life that evolved *electron transport chains* from which photosynthesis later developed.

CHLOROPHYLL: from the Greek **khloros** 'green' + **phullon** 'leaf', is a *tetrapyrrole* or *porphyrin* compound similar to that of haem but containing a magnesium ion coordinated to the four pyrrole nitrogens.

CHLOROPLAST: **khloros** + **plastos** 'formed' is the *organelle* or *plastid* which is the site of photosynthesis in plants. Photosynthesis relies on the absorption of light photons by *chlorophyll* located in chloroplasts. In plants, as distinct from photosynthetic bacteria, photosynthesis consists of a coupled system composed of *photosystem I* and *photosystem II* (PS I and II). Both systems absorb light by chlorophyll in membrane-bound structures known as a *thylakoids*. PS II is responsible for the decomposition of water with the liberation of oxygen. The protons released in this reaction generate a proton gradient across the thylakoid membrane which drives ATP synthesis. It also supplies electrons to PS I, which boosts them to an energy high enough to reduce *NADP⁺*. Electrons flow from PS II to PS I via a complex known as *cytochrome bf*. This complex corresponds to complex III in mitochondria and most components of cytochrome bf are homologous to those of complex III. The ATP synthases in both systems are also similar. The term *thylakoid* comes from the Greek **thulakos** 'pouch', which also is the origin of the common name *thylacine* for the extinct Tasmanian tiger, *Thylacinus cynocephalus*, referring to its marsupial pouch. Thylakoid membranes derive from the inner chloroplast membrane and so correspond to *cristae* in mitochondria.

NADPH, the final reduction product of photosynthesis, generally participates in synthetic reactions, while NADH is normally the product of degradative reactions. The so-called dark reactions of photosynthesis begin with formation of a 6-carbon sugar by carboxylation of ribulose 1,5-bisphosphate by *ribulose bisphosphate carboxylase*, also known as *rubisco*. Rubisco, being a relatively slow enzyme, accounts for 50 per cent of the total protein in chloroplasts and is estimated to be the most abundant protein on earth.[84] Net synthesis of sugar then occurs by successive rounds of the Calvin cycle utilizing the rubisco product, NADPH and ATP. See also entry for **CYCLE**.

PLASMA (Gk.): anything formed, shaped or moulded which, for the Greeks, meant anything made from wax or clay. The corresponding verb, 'to mould' was **plassein** while the adjective **plastikos,** meaning 'pliable', is the origin of English 'plastic'. Numerous words in English are based on the **plasma** root, but its meaning in English is probably best thought of as 'substance'.

Blood plasma: colourless fluid remaining if cells and fat globules are removed.

Cytoplasm: the material in a living cell, excluding the nucleus.

Cytoplasmic membrane: the external membrane as distinct from other membranous material in the cytoplasm.

Plasmalemma: a term applied to the cytoplasmic membranes of plant cells. In muscle cells, it is *sarcolemma*: **sarx, sark-** (Gk.) 'flesh'; **lemma**: husk or rind.

Neoplasm: abnormal growth of tissue; **neo-** (Gk.) 'new', while *neoplasia* refers to the presence or development of a neoplasm. Similarly, *dysplasia*: abnormal cell or tissue growth and *metaplasia*: the transformation of one type of differentiated tissue to another type.

Mycoplasma: the group of bacteria misnamed because they look like fungi.

Protoplast: a bacterial cell from which the cell wall has been removed.

Plastids: organelles such as *chloroplasts* in plants.

Also, *Apicoplast*—see entry above for **APICOMPLEXA** and likewise for **ENDOPLASMIC RETICULUM.**

PLOIDY: the number of sets of chromosomes in a cell.

HAPLOID: **haploos** (Gk.) 'single' as applied to germ cells, sperm, and oocytes that contain a single set of chromosomes. The related term HAPLOTYPE is used to describe sequence variants, usually single nucleotide polymorphisms (SNPs), of a gene or a group of genes on a single chromosome; it can also refer to particular HLA variants—see **IMMUNITY.**

DIPLOID: **diplous** 'double', describes the standard somatic cell containing two sets of chromosomes, one from each parent.

TRIPLOID, TETRAPLOID, HEXAPLOID: cells containing three, four, or six sets of chromosomes, not uncommon in plants. ANEUPLOID: **a** 'not' + **eu** 'true' + **ploid**: cells containing fragmented or translocated chromosomes or simply an abnormal number of chromosomes. Tumour cells are often aneuploid. Mammalian liver also displays unusual ploidy in that a small but variable proportion of cells are observed to contain two nuclei while both mononucleated and binucleated hepatocytes can be polyploid, containing 4x, 8x, 16x, and higher the normal haploid DNA content. Studies in mice show that ploidy in hepatocytes can increase as a result of failed cytokinesis or it can decrease as a result of mitoses occurring in the absence of DNA replication. This latter process can result in aneuploidy as well as cells containing uniparental chromosome sets. The suggestion is that such variation in gene dosage allows cells to adapt to nutritional or xenobiotic stress.

COLCHICINE is a toxic alkaloid obtained from the autumn crocus *Colchicum autumnale*. It is prescribed for gout. Scientifically, its interest lies in its ability to bind to the protein *tubulin*, and thereby to prevent its polymerization to form *microtubules* which function to separate homologous chromosomes during mitosis and meiosis. When colchicine is applied to cells undergoing meiosis, the result is the formation of gametes either with no chromosomes or with double the normal number. This procedure has been used on plants to generate tetraploid and other polyploid varieties which often have improved properties. Triploids can be obtained by crossing diploids and tetraploids. Colchicine is also being investigated as a component of an anticancer formulation.[85]

PNEUMO-: relating to the lungs, **pneumon**, from **pneuma** (Gk.) 'wind or that which is breathed or blown'. *Pneumonia* is a disease of the lungs, of *pneumocytes*, which can be caused by bacteria, viruses, or an autoimmune condition and is characterized by inflammation of the small air sacs of the lung known as *alveoli*, **alveolus** (L.), the diminutive of **alveus** 'hollow or cavity'. The corresponding verb 'to breathe' is **pnein** (Gk.), and the condition associated with temporary cessation of breathing is known as *sleep apnoea*, from **apnous** 'breathless'. *Pneumothorax* is another name for 'collapsed lung', the situation where air is present between the lung and the chest wall. *Pneumatic tools* are those operated by compressed air such as jackhammers and drills. There was a time when motor car tyres were referred to as *pneumatic tyres*, presumably to distinguish them from solid rubber tyres. The Latin for lung(s) is **pulmo, pulmon**-, from which we get *pulmonary artery*, *pulmonary embolism*, etc. The Latin equivalent of **pnein** is **spirare** 'to breathe', which gives rise to *respire* and *respiratory*, and from the same root, *conspire, expire, inspire*.

PROTEINS are macromolecules in which the repeating units are *amino acids* joined together by *peptide bonds*, also referred to as amide bonds, formed by the successive reaction of the *carboxyl group* of the first amino acid with the *amino group* of the next amino acid. The first amino acid of the final protein will thus have a free amino group while the final amino acid will have a free carboxyl group, and these amino acids are referred to as the *N*- and *C-terminal* amino acids respectively.

THE HUMAN PROTEOME: this expression refers to the total complement of proteins encoded by the human genome which are synthesized at different times and in different human tissues. Completion of the Human Genome Project allowed an estimate to be made of the number of protein-coding genes by a process of *annotation* of the genome, involving a search for *open reading frames* or ORFs. This is not straightforward because most eukaryotic individual protein coding sequences consist of a series of coding segments, the *exons*, interrupted by non-coding *introns*. One way round this is to isolate *messenger RNAs* (mRNAs) from which the introns have been *spliced* out and the exons joined together in the final mRNA. Isolation of total mRNA from a *unicellular* organism such as yeast corresponds then to the yeast *exome*. But for *metazoans* such as us, different tissues synthesize different proteins and so different mRNA populations. It is now possible to extend this type of analysis to the single-cell level.

The total human proteome consists of about 20,000 proteins. So far there is direct protein evidence for ~17,000 proteins and RNA evidence for a further ~2,500 and another ~700 based on annotation for which there is so far neither protein nor RNA evidence. In a survey of forty-four different tissues, ~9,000 genes were found to be expressed in all tissues. These correspond to what are called *housekeeping* genes that specify metabolic enzymes, cytoskeletal and other structural proteins. Thirty-four per cent of genes show elevated expression in at least one tissue, the largest numbers of such genes being expressed in testis, brain, and liver. About 5,500 proteins are predicted to be *membrane-bound*, and about 3,000 are predicted to be *secreted proteins*, termed the *secretome*.[86]

No fewer than 72 per cent of genes give rise to multiple *splice variants*, or *isoforms*, as a result of *alternative splicing* in which some exons are excluded during the splicing process that forms mRNA from the *primary transcript*. Other sources of variation include enzyme-mediated *post-translational modification* such as phosphorylation, acetylation, methylation, proteolytic cleavage, as well as genetic variation. The addition of hydrophobic *myristoyl* or *farnesyl* groups has the effect of causing proteins to associate with membranes. Proteins destined to be secreted are synthesized with additional *leader peptide sequences* at their N-terminus which are subsequently removed after secretion. Many enzymes, particularly proteases, are synthesized as inactive precursors called *zymogens* or *proenzymes*, which are later activated by limited proteolysis, e.g. trypsinogen, proelastase. Numerous examples occur in the zymogen activation cascade that occurs during blood clotting. See also **BLOOD** entry. Enzyme precursors that are synthesized with a leader peptide and which require activation are referred to as *preproenzymes*.

PROTEIN STRUCTURE: A major division of proteins according to structure is between proteins which in their final folded state are reasonably *compact*, or *globular*, as against proteins which are *fibrous* proteins or which contain a mixture of fibrous and globular parts. Many of the latter are very large proteins and often play a structural role. Protein structure is conveniently considered under the headings primary, secondary, tertiary, and quaternary structure. The *primary structure* of a protein is defined as its amino acid sequence, while *secondary structure* corresponds to *regular* 3D structures that include the α-*helix*, the β-*pleated sheet*, and the *reverse turn*. Each of these features are stabilized by the formation of *hydrogen bonds* between the C=O group of one amino acid and the N-H group of another. For the α-helix, adjacent amino acids are separated by a 100° turn of the helix and a vertical displacement of 1.5 A. There are thus 3.6 amino acid residues per turn of the helix, and the *pitch* of the helix, or its *repeat*, equals 3.6×1.5 A $= 5.4$ A. The β-sheet results from hydrogen bonds forming between two or more separate, fully extended *polypeptide strands* which may extend, either in opposite directions, i.e. *antiparallel*, or in the same direction, to give a *parallel* β-sheet. Reverse turns occur when a strand reverses direction and involves hydrogen bonding between the C=O of one amino acid and the N-H group of the third amino acid further along the chain. Protein *tertiary structure* consists of the final, folded, 3D structure of a protein, while *quaternary structure* refers to protein complexes with more than one component. Examples are the $\alpha_2\beta_2$ structure of haemoglobin and the trimeric αβγ G proteins that respond to G protein-coupled receptors (GPCRs), see **RECEPTORS**.

PROTEIN DOMAINS: These are sequences of amino acids able to fold independently to form a compact 3D structure. Folding involves sequestering most non-polar residues in the *interior* of the molecule with polar and charged residues mainly on the *exterior* in contact with the aqueous environment. Small globular proteins consisting of less than 200 amino acids, such as the enzymes ribonuclease or lysozyme, consist of a single domain. Large proteins usually contain multiple domains, and it is estimated that 65 per cent of eukaryotic proteins are *multi-domain*. The same domains are often found in different proteins joined by flexible *linkers*, typical examples occurring in proteins of the extracellular matrix and cell surface adhesion molecules. There is a lower limit on the size of protein domains. In the absence of disulphide bonds, sequences of fewer than about forty amino acid residues are unable to form a stable fold.

PROTEIN FOLDING: An important early experimental result was that the folded tertiary structure of a protein depends on its amino acid sequence alone and that no additional information is required for a protein to fold into its final shape or conformation. These experiments consisted of subjecting purified enzymes to denaturing conditions in concentrated solutions of urea together with reagents that broke disulphide bonds. When these were removed and the protein incubated under renaturing conditions, the enzymatic activity returned. Subsequently it has become apparent that protein folding *in vivo* is much more challenging and that there are a large number of proteins termed *chaperones*, whose function in the cell is to assist protein folding.

MOLECULAR CHAPERONES: As noted above, renaturation experiments *in vitro* led to the conclusion that the amino acid sequence is sufficient to determine protein tertiary structure. For *nascent* (L. **nasci** 'to be produced') proteins *in vivo*, the situation is very different, and the term *nascent* implies (particularly for large proteins) that they may need to begin folding while translation is still in progress, i.e. initial folding may occur *cotranslationally*. These emerge from the ribosome into the *cytosol* of the cell which contains 10–20,000 different proteins that together account for an extremely high total protein concentration that favours *aggregation*. Folding involves sequestering hydrophobic sequences out of contact with water and away from similar sequences that would otherwise result either in *misfolding* or intermolecular aggregation. The problem is solved by the presence of proteins termed *molecular chaperones* which, by associating with the hydrophobic sequences of unfolded proteins, prevent aggregation reactions and assist folding in various ways as well as targeting the final products to defined locations such as the endoplasmic reticulum or mitochondria.[87]

PROTEIN DEGRADATION: All proteins turn over and do so at varying rates. One estimate is that 2×10^{11} human cells die every day. It is clear that methods are required for ridding cells of used or damaged proteins. One of two main methods for doing this consists of *autophagy*—see under **PHAGE** entry for details. This is also thought to take care of aggregates forming as a result of unfolding or misfolding of proteins during synthesis. Such proteins are accumulated within *aggresomes*, which are cleared via autophagy which culminates in proteolysis in *lysosomes*.

The other method of protein disposal depends on *ligation* to the small protein *ubiquitin*, which then delivers it to another large complex, the *26S proteasome*, the large value for its *sedimentation coefficient*, S, being an indication of its size. This system, known as the *UPS system*, has the advantage that it is capable of degrading *specific* proteins. This comes about because there are literally hundreds of different enzymes that carry out the *ligation* reaction enabling specific *ubiquitination*. An example of where this specificity is crucial involves proteins that carry out specific steps in the closely regulated *cell cycle* but which then need to be degraded. Ubiquitin is found in all cells, hence its name.

PROTEOSTASIS is really an abbreviation of *protein homeostasis* and can refer to the whole gamut of protein reactions from synthesis, folding, targeting, maintenance, and degradation, all of which need to be carefully regulated to maintain proteostasis. The effects are apparent when it is not maintained as a result of old age or the ravages of disease.

PYR-, PYRO- (Gk.): 'fire'. In the early days of chemistry, many new compounds were discovered by the destructive distillation of other compounds. Many of these were given names with the prefix *pyro-*. A neutral substance obtained by the distillation of calcium acetate was described as 'pyro-acetic spirit'. This was later renamed *acetone*. Pyruvic acid (**pyr-** + **uva** 'grape') was obtained by the distillation of tartaric acid obtained from grapes. Heating phosphoric acid gives rise to *pyrophosphoric acid*, $2H_3PO_4 \rightarrow H_4P_2O_7 + H_2O$. Numerous current names of chemicals indicate their descent from chemicals named in the early 1800s: *pyrrole*, *pyridoxal*, *pyrimidine*. Related words include *pyromaniac*, *funeral pyre*, *pyrolysis*.[88]

RECEPTORS are ubiquitous since they are the means by which signals, as distinct from molecules, are transmitted from outside the cell to its interior. Most receptors are *membrane-bound proteins* that are in contact with both the extra- and intracellular environments. A small number of receptors act in the nucleus—see below. Receptor function depends on specific binding of a *ligand* or *agonist* to the extracellular portion that has the effect of inducing a conformational change in the receptor which is transmitted to the interior of the cell that often then sets in motion a *signalling cascade*. Ligands can be *hormones*, *neurotransmitters*, or *peptides*. Such receptors are said to be *ligand-gated*. There are other receptors which control membrane *ion channels* that are *voltage-gated*. Receptors can be classified into families depending on their structure, function, and pharmacology. The polypeptide chains of most receptors pass backwards and forwards through the membrane multiple times. These *transmembrane* regions, or TMs, are recognizable because they contain a run of hydrophobic amino acids. In many cases, it has been shown that ligand-induced *conformational changes* consist of alterations to the interactions of these TM regions, which in turn induce a cascade of intracellular molecular interactions that determine gene expression.

G PROTEIN-COUPLED RECEPTORS: Determination of the human genome sequence revealed the presence of over 800 genes coding for G protein-coupled receptors (GPCRs). The transmembrane region of GPCRs is relatively conserved and consists of a characteristic bundle of *7 alpha helical regions*, referred to as 7TM. The extracellular region contains the receptor N-terminus and three extracellular loops which together form a pocket for binding of a ligand or *agonist*. It is the precise architecture of the binding pocket that determines which ligand will bind and the strength of binding. The cytoplasmic side likewise contains three loop structures, the C-terminus, and the site of binding of *heterotrimeric* G proteins (Gαβγ) or other intracellular signalling proteins. They respond to a variety of signals, including photons which cause isomerization of *11-cis retinal*, the *prosthetic group* of the photoreceptors—see under **OPSINS** above. The 7TM receptors also mediate numerous other biological functions including smell, taste, and development. Receptors responding to light, odour, and taste are referred to as *sensory GPCRs*.

Drugs that target these receptors are either already being used or being tested for conditions that include abnormal heart rate, asthma, Parkinson's disease, and schizophrenia. A group of four receptors bind histamine. Histamine is released from *mast cells* which are stimulated by various antigens, giving rise to allergic reactions and inflammation. Much research has gone into finding suitable *antihistamines*, these being ligands that bind to the receptors and block histamine binding. These ligands are termed *antagonists*, they block activation of the receptor and prevent or alleviate conditions such as asthma. The outstanding success of the antihistamine work suggests that other GPCR-related conditions can be made to yield to further research.

OPIOID RECEPTORS: **opion** (Gk.) 'poppy juice'. Painkillers such as *morphine* and 3-methyl morphine or *codeine*, from **kodeia**, 'poppy head', bind to opioid receptors which are widely distributed in the brain and peripheral nervous system. Opioid receptors are also members of the GPCR superfamily. The binding of a ligand such as *morphine* to its GPCR results in generation of an intracellular signal mediated by the trimeric G protein. Detailed x-ray crystallographic structures are now available for all four opioid receptors, and these may enable the design of ligands that retain the painkilling properties of morphine and other opioids but without undesirable addictive side effects which in recent years have led to a lethal epidemic of opioid overdosing.

ENDORPHINS: an abbreviation of endogenous morphine, **endo-** + **morphin**, after Morpheus, the Greek god of dreams, and son of Hypnos, the god of sleep. β-endorphin is a peptide thirty-one amino acids in length which is generated by the proteolytic processing of *proopiomelanocortin*. It is found in both the central and peripheral nervous system. It binds to the μ-opioid receptor, the receptor to which morphine, the analgesic drug obtained from opium poppies, binds and is thought to play an important role in determining aspects of behaviour and responses to pain and stress. See also ENKEPHALINS under **CEPHALO-**.

SENSORY RECEPTORS: a distinct family of cation channels are the *transient receptor potential* (TRP) ion channels. These are *sensory receptors* which respond to a wide range of chemical and physical stimuli including heat, cold, pH, and physical forces. Unlike receptors described above, they are neither ligand-gated nor voltage-gated. There are twenty-seven members of this family in humans. A subfamily of these are heat sensors, known as TRPV, where V stands for *vanilloid* because the first of these to be characterized, TRPV1, is also a receptor for *capsaicin*, a derivative of vanilla and one of several related compounds derived from *chilli peppers*. The TRPV1 channel normally opens between 37° and 43°C, resulting in an influx of Ca^{++} which sets in train a neuronal response that registers as heat. Capsaicin binding to TRPV1 also causes the channel to open, producing a burning sensation. Binding of toxins from spiders and plants may also result in a pain response. Not all receptors are 7TM. The TRPV1 receptor and the acetylcholine receptor, both of which regulate ion channels, consist of four subunits, each with 6TM regions.

NUCLEAR RECEPTORS: Nuclear receptors bind ligands, which are small hydrophobic molecules able to diffuse through cell membranes, and include cholesterol, androgens, oestrogens, and cortisol as well as fat-soluble vitamins A and D. The human genome sequence indicates a total of forty-eight evolutionarily related genes encoding nuclear receptors. As well as a DNA binding site, each receptor has a relatively conserved ligand-binding site or *domain* which is constructed from up to twelve α-helices. These receptors also include so-called *orphan receptors* for which no ligand has yet been identified. The ligands bind to the receptors in the cytoplasm, the resulting complexes migrate to the nucleus, where they bind to DNA, acting as *transcription factors*.

SALT: halos (Gk.) and **sal** (L.), hence *halogen*, also *saline* as in *physiological saline*. The term *halogen* indicates their ability to form salts such as sodium chloride. The halogens, Group VII in the periodic table, are fluorine from **fluere** (see above), chlorine from **chloros** (Gk. 'green', named on account of its colour), bromine from **bromos** (Gk. 'stench'), iodine from **iodes** (Gk. 'violet-coloured'), astatine from **astatos** (Gk. 'unstable', because it really is unstable, being radioactive). A *halophile* is an organism such as a bacterium that can grow in high levels of salt, and plants with this ability are *halophytes*.

SCIENTIFIC KNOWLEDGE: THEORY and PRACTICE

Scientific knowledge accumulates by the invention of hypotheses and theories, which are then tested by experiment; i.e. *theories* are supported or disproved by *practice*. In the academic context, theory and practice correspond to lectures and practical classes, or what Bertrand Russell termed 'knowledge by description and knowledge by acquaintance'. The desirability of combining theory and practice has long been appreciated in artistic circles. A quote in the *OED* from 1795 says, 'To be learned in an art etc. the theory is sufficient; to be a master of it both the theory and practice are requisite.'

THEORY comes from **theoria** in both Latin and Greek, where it meant 'contemplation' or 'speculation'. Its definition in English, from *OED2*: 'A scheme or system of ideas or statement held as an explanation or account of a group of facts or phenomena, *but also a hypothesis that has been confirmed or established by observation or experiment*'. The second part of this definition needs to be remembered when creationists declare that evolution is only a theory. The historian Tony Judt reported, 'I knew a man in California whose doctoral dissertation was devoted to 'Theory and Practice in theory and practice'. PRACTICE is from **praxis** (Gk.) meaning 'practice', also 'practical ability', and from the Greek **techne** meaning art or craftsmanship and giving rise to *technique*, *technical*, and related words that pertain to practical ability.

EXPERIMENT comes directly from Latin **experimentum**, which meant 'test' or 'trial', derived in turn from the verb **experior** 'to try', from which we also get *expert*, *expertise*, and *experience*. Another word is **experientia**, meaning 'knowledge gained by trial'. There used to be a scientific journal in English called *Experientia*. It has now been rebadged as *Cellular and Molecular Life Sciences*, which could be taken to illustrate the point that Latin, compared to English, is a succinct language. The Latin suffix -**entia** also occurs in the word **scientia** meaning 'knowledge' and which comes from **scire** (L.) 'to know' and from which we get the word *science*. The Greek word for knowledge, **gnosis**, is related to the Latin **cognoscere** 'to get to know, learn', and from one or the other, we get *prognosis*, *recognition*, *diagnosis*, *agnostic*, *cognitive*.

The word *science* has been in use since Chaucer's time and, earlier on, was applied to any area of learning or expertise, from theology to law and the arts. Curiously, it was not until 1833 that the word *scientist* was coined to describe people who did what the poet Coleridge described at the time as 'real science'. The need for a term to describe people so employed was debated at the 1833 meeting of the British Association for the Advancement of Science, and Whewell suggested that by analogy with the term *artist*, the word *scientist* should be coined. He went on to argue that such a word should be acceptable because words such as *economist* and *atheist* already existed. In a case of guilt by association, mention of the word *atheist* resulted in an outburst by Adam Sedgwick: 'Better die of this want [of a term] than bestialize our tongue by such a barbarism'.[89, 90, 91] In spite of Sedgwick, *scientist* soon caught on.

EPISTEMOLOGY, or the theory of knowledge, from **episteme** (Gk.) meaning 'knowledge', relates to how scientific knowledge is obtained and how science progresses. The traditional view, which goes all the way back to Francis Bacon (1561–1626), holds that science proceeds by way of an accumulation of observations in a particular area that eventually enables a *general* conclusion or statement to be made. This procedure is known as *induction*. The well-worn illustration of this approach says that if all swans that have ever been observed are white, then eventually, we can make the general statement 'All swans are white.' David Hume (1711–1776) pointed out that repeated consistent observations cannot *logically* prove a proposition, i.e. in this case, it could not exclude the possibility that one day someone somewhere would observe a black swan. This led to the embarrassing conclusion that the whole of science rested on shaky foundations, although it has to be said that this caused more sleepless nights among philosophers than it did among scientists. Fittingly, it was the philosopher Karl Popper (1902–1994) who solved the problem by proposing that science progresses not by attempting to *prove* propositions, hypotheses, and theories but by all-out attempts to *disprove* them. Hence, observation of a single black swan leads to the conclusion 'Not all swans are white.' Clearly, a good working hypothesis must be non-trivial and amenable to experimental investigation. Most scientists agree that this really is how science progresses.

-SCOPE: **skopein** (Gk.) 'to look at'. When the Greeks made observations, they could use only their eyes. Progress in science and medicine has depended on the invention of equipment and procedures for observing things that are not visible to the naked eye. The *microscope* revealed the presence of bacteria, and electron microscopes reveal structures at the atomic level. Galileo used the *telescope*—**telos** (Gk.) 'at a distance'—to discover the moons of Jupiter, while the Hubble telescope above the obscuring atmosphere gathers light from galaxies that formed not long after the big bang. The gold-leaf *electroscope* was used to demonstrate the presence of electrical charge, while cathode ray *oscilloscopes* provide a read-out of varying voltages and can be used to monitor varying non-electrical signals such as sound after their conversion to voltages. Many different applications of *spectroscopy* are used to study the interaction of light and matter. During the nineteenth century, a number of elements were discovered by observing their emission spectra following thermal excitation—see for example the entry for **HELIUM**. The *stethoscope*—**stethos** (Gk.) 'chest'—is used to assess heart and lung function, while *endoscopy* is a procedure for monitoring internal organs. Greek **episkopos** means 'overseer', but in English, *episcopal* refers to a particular person who oversees, namely a bishop.

SECRETION MECHANISMS

ENDOCRINE: **endo-** 'inside, within' and EXOCRINE: **exo-** 'out' + **krinein** 'to sift or separate' but best understood here as meaning 'to secrete'. The endocrine system consists of glands which secrete hormones directly into the bloodstream. Compared to the nervous system, responses in the endocrine system are much slower but longer-lasting. The endocrine glands located in the brain, i.e. the *hypothalamus*, *pituitary*, and *pineal* glands, are together responsible for secretion of about twenty different hormones. In descending (anatomical) order, the other endocrine glands are the *thyroid* and *parathyroid* glands, the *adrenals*, the *pancreas*, *ovaries* and *testes*. The endocrine system is distinguished from the EXOCRINE system, which includes *sweat glands*, *salivary glands*, *mammary glands*, *gastrointestinal glands* (*mucin*), and the liver (*bile*), where secretion occurs via a system of ducts.

AUTOCRINE: **autos** (Gk.) 'self, directed from within'. Growth factors, in this case autocrine factors, exert their effect in an autocrine manner when they act back on the cells that secrete them. This means that these cells also express surface receptor molecules that are specific for the secreted factor. An example is *vascular endothelial growth factor* (VEGF) that is secreted by stem cells and which binds to a receptor on these same cells and stimulates their proliferation in an autocrine manner. It has become apparent that VEGF also acts in a PARACRINE manner (**para-** here in the sense of 'beside' or 'adjacent to'). Paracrine factors are those that act on cells in the *vicinity* of the cell secreting it, in contrast to the endocrine system. In the case of VEGF, it acts on the endothelial cells that line blood vessels and stimulates their proliferation. The result, *angiogenesis*, is blood vessel formation (**angeion** Gk., 'vessel'). VEGF and other growth factors involved in angiogenesis are also described as *angiocrine* factors.

APOCRINE: **apokrinein** (Gk.) 'to set apart', refers to a particular mechanism of exocrine secretion in which material is secreted within a membranous vesicle because of portion of the secreting-cell membrane budding off. Fat droplets in milk are secreted in this fashion. A type of *sweat cell* is described as apocrine. These are restricted to armpits and pubic areas, i.e. hairy areas where the secretion occurs into the top of the hair follicle. Because they contain protein which is broken down by bacteria, these apocrine secretions are responsible for body odour. The other type of sweat cell is described as ECCRINE (**ek-** 'out + **krinein**), in which sweat secretion occurs without loss of cell material. These eccrine glands are spread over most of the human body, and their function is to cool the body as the sweat evaporates.

EXOCYTOSIS: Some exocrine secretions employ a different mechanism in which a vesicle containing the material to be secreted merges with the cell membrane and empties the contents of the vesicle into the exterior. An example is the release of acetylcholine into the synaptic cleft, triggered by the arrival of a nerve impulse, in which the membranous vesicles that stored the acetylcholine in the presynaptic axon become part of the presynaptic membrane.

ENDOCYTOSIS: can be regarded as the opposite of exocytosis, whereby compounds unable to pass through cell membranes are engulfed by a portion of the membrane and transferred as vesicles to the interior of the cell. This process obviously consumes portion of the cell membrane and it is suggested that, depending on the type of cell, this can be compensated for by exocytosis.

SELENIUM: selene (Gk.) 'moon', atomic number 34, is in group VI under sulphur in the periodic table. It is an essential *trace element* in many organisms while around 50 µg/day is essential for human health. Selenium deficiency, although not often encountered in humans, has been identified as a contributing factor in a range of pathologies. Selenium was discovered and named by Berzelius in 1817, who found it associated with the element *tellurium*, atomic number 52, also group VI, which had been named after the Latin word **tellus** 'earth'. It occurs in a small number of proteins as an unusual amino acid, *selenocysteine* (Sec), in which the thiol (-SH) of cysteine is replaced by a *selenol* (-SeH) group, and is the twenty-first amino acid occurring naturally in proteins. It is coded for by the codon UGA, which normally is a chain termination codon—see Genetic Code table. Sec is inserted into proteins when an in-frame UGA sequence occurs together with a *stem-loop* structure, a *cis*-acting *Sec insertion sequence* (SECIS), located downstream in the 3' non-translated region of the mRNA. Proteins containing Sec residues are found in bacteria, archaea, and eukaryotes, indicating an ancient origin.

Twenty-five selenoproteins, also referred to as the *selenoproteome*, occur in humans. The functions of several of these is not known; however, most are enzymes involved in *oxidation-reduction* reactions. The number of Sec residues in these proteins varies between 1 and 10. A Sec residue is part of the *active sites* of these enzymes, which include five different *glutathione peroxidases* (GPx). See under **ANTIOXIDANTS**. Another group of Sec-containing enzymes are involved in modifications of the thyroid hormone *thyroxine*. Thyroxine, T4, contains four iodines and selenoproteins are involved in reductive removal of one iodine to form T3, *triiodothyronine*, which is more active than T4. Another group of selenoproteins are the *thioredoxin reductases*, which play an important role in cellular redox homeostasis.[92]

SENESCENCE: senescere (L.) 'to grow old', a general term to describe the ageing of organisms. More recently, senescence at the cellular level has been investigated. Senescent cells are those that have permanently stopped dividing because of disease or injury or other reasons but remain metabolically active and typically accumulate in old age. The mechanisms that induce senescence are not completely understood. Senescent cells are thought to play a role in tissue remodelling during embryogenesis and wound healing and to protect against cancer in young animals. In old age, senescent cells are associated with sites affected by diseases such as *osteoarthritis* and *atherosclerosis*. Such cells secrete a cocktail of compounds including cytokines, growth factors, and proteases, designated the *senescence-associated secretory phenotype* (SASP), with deleterious effects on surrounding cells and, in some situations, creation of a tumour-permissive microenvironment.

These observations have suggested experiments to examine the effects of treatments that remove senescent cells. Experiments of this sort have so far been carried out on mice treated with compounds known as *senolytics*, which remove senescent cells. Senescent cells persist and accumulate because they are resistant to *apoptosis*, while senolytics overcome this resistance. Senescent cells in different tissues respond to different senolytics. In experiments with mice, senolytics have been observed to delay neurologic dysfunction, to relieve pulmonary fibrosis that is senescence-driven, to improve gait, and to decrease bone loss and arteriosclerosis in old animals. In confirmation of previous work that pointed to a role for senescent cells in ageing, senolytics have extended the lifespan of mice by up to 36 per cent. All of this work so far has been on mice, and human trials are just beginning.[93]

SEQUENCE: sequens 'next or next following' from **sequor, sequi, secutus** (L.) 'to follow'. *Sequence* as a noun simply means the order in which the monomer units of a polymer are arranged; as a verb, to sequence means to *determine* the order of monomer units in a polymer, e.g. the sequence of amino acids in a protein or of nucleotides in a polynucleotide. In principle, knowledge of the genetic code means that determination of a DNA sequence also provides the sequence of the protein coded for by that sequence. In practice, this requires additional information about the location of the first amino acid, and in eukaryotes, the position of *introns* and knowledge of the size of the protein can also be useful. Deduction of the proteins encoded by a genome is referred to as *annotation* and, with only the DNA sequence to go on, is not a straightforward process. Knowing the sequence of the corresponding mRNA helps greatly, particularly with respect to the positions of introns, though even here, further complications can arise because of *alternative splicing*. Related words from the same root are *consecutive, consequence, sequential, persecute, obsequious, prosecute, sequel*. Developments in DNA sequencing techniques since the original methods in 1977 have seen the cost of sequencing the human genome fall from the many tens of millions of dollars that the original Human Genome cost by the time it was finished in 2003 to about $1,500 by 2015.[94]

SLEEP (OE). According to the *OED*, first recorded as both noun and verb ca. 850; also *slumber*, another OE word but dating from the 1300s. Hypnos was the name of the god of sleep in Greek mythology, and **hypnoin** (Gk.) meant 'to put to sleep'. English has the combining form *hypno-* as in *hypnosis*, defined as artificially induced sleep, hence *hypnotize, hypnotist, hypnotherapy*. The Romans had **somnus** for 'sleep', from which we get *somnolent* and *insomnia*, which was also their word for sleeplessness. Another of their words for sleep was **sopor** meaning 'deep sleep' but sometimes with suggestions of pathology or permanence. One associates the English *soporific* with TV programs. *Coma* from **koma** meant 'deep sleep' in Greek and lack of *consciousness* in English.

NARCOLEPSY: a condition characterized by an extreme tendency to fall asleep, from **narkoun** (Gk.) 'to make numb'—the root for *narcosis* and *narcotics*, while *-lepsy* derives from **lambanein** 'to take hold of', the same root as in *epilepsy* and *catalepsy*, conditions that can involve seizures or convulsions of varying intensity. Narcolepsy can therefore be described as an 'attack of sleep'. Narcolepsy affects about 1 in 2,000 people. It is described as a *spectrum disorder* because in different people it may entail a variety of symptoms of variable intensity. Typically, it involves the sudden onset of sleep and/or complete loss of muscle tone, which goes by the name of *cataplexy* from **cata-** 'down'+ **plessein** 'to strike', literally 'struck down'. Attacks are short-lived and commonly brought on by emotional triggers such as laughter, surprise, or anger. Most cases of narcolepsy are due to the autoimmune destruction of a small group of cells in the hypothalamus. These cells produce a pair of neuropeptides called either *hypocretins* or *orexins*.[95],[96]

FATAL FAMILIAL INSOMNIA: a very rare late-onset, autosomal dominant condition, which means it will be transmitted to 50 per cent of the children of a person carrying the mutation. A few sporadic cases have also been diagnosed. FFI patients invariably have mutations in the gene that codes for the *prion protein PrP*. This is the protein that gives rise to *amyloid plaque* formation, characteristic of neurodegenerative diseases such as Alzheimer's. Amyloid formation can occur with a range of different proteins and is generally regarded as the result of protein unfolding or misfolding. The association of FFI with amyloid formation has been interpreted as consistent with the idea that the function of sleep is to clear the brain of molecular garbage.

SODIUM and POTASSIUM. These two metals were isolated by Humphry Davy in 1807 by electrolysis of solutions of *caustic soda* and *caustic potash* derived from what for centuries had been called *soda ash* and *potash*. The traditional source of soda ash was a *halophytic plant* growing in coastal regions of the Mediterranean that sequesters sodium chloride in large *vacuoles* that can occupy up to 80 per cent of cell volume. Soda ash was prepared by burning the leaves and extracting the ash with water. Potash, *potasse* in French, was prepared in a similar manner from the ash of hardwood trees, and named for the pots in which extracts were concentrated. The aqueous extracts of soda ash and potash consisted mainly of the weakly alkaline carbonates. Heating to drive off carbon dioxide and addition of water produced the strongly alkaline hydroxides, caustic soda, and caustic potash (**kaustikos** Gk. 'burning').[97]

THE SODIUM-POTASSIUM PUMP. Cells of higher organisms have levels of intracellular potassium that are high compared to those of sodium. The opposite situation prevails on the extracellular side. This is maintained because of activity of the *sodium-potassium pump*, which consists of a transmembrane enzyme known as the Na^+/K^+ ATPase, which for each cycle exports three Na^+ while importing two K^+. ATP hydrolysis is involved because both Na^+ and K^+ are being transported against *concentration gradients*. A consequence is that the inside of the cell membrane bears a net negative charge and the outside a net positive charge which manifests as a *potential difference* across the membrane of about -70 millivolts. The membrane is said to be polarized, and this is the case for most cells.

ACTION POTENTIALS. In nerve tissue, i.e. *neurons*, the transmission of nerve impulses occurs via the sequential *depolarization* of the *axon* membrane. This is referred to as an action potential, which consists of the transient sequential opening and closing of *voltage-gated* sodium and potassium *ion channels* along the axon. The process initiates with release by the *presynaptic neuron* of a *neurotransmitter*, which binds to a *receptor* on the *postsynaptic neuron*. This has the local effect of causing a slight drop in the resting membrane potential, which in turn causes a voltage-gated sodium-ion channel to open, an influx of Na^+ ensues, and the membrane potential goes to +40 millivolts. Within about one millisecond, the sodium channel closes and a potassium channel then briefly opens, which rapidly restores the local membrane potential. As successive action potentials are triggered along the axon, the nerve impulse proceeds either to another neuron or perhaps a *neuromuscular junction*.

SOLVENT derives from **solvo, solvere, solutum** (L.), where the meaning was 'to loosen, unfasten'. Related words are *solute, solution, soluble, dissolve, dissolute, resolute*, and *solve*. The English word *absolve* has the meaning 'to exempt from obligation or blame', while *absolute* shows up in English expressions such as *absolute power* and an *absolute majority* in parliament as well as *absolute alcohol* and the *absolute temperature scale*. The Greek word corresponding to *solvere* was **luein**—see entry for **LYSIS**.

SOMATO- **somatikos** 'of the body', from **soma** (Gk.) 'body'.

SOMATIC CELL: any cell of an organism apart from its germ cells

SOMATOSTATIN: **soma** + **statum** (L.) from **stare, steti, statum** 'to halt or arrest'. It is synthesized in the *hypothalamus* as a precursor which is cleaved into a 14-residue peptide as well as an extended 28-residue species. It inhibits release of *growth hormone* from the anterior pituitary. Release of growth hormone is also dependent on the presence of *growth hormone releasing factor* (GRF), the absence of which can lead to a form of *dwarfism*. It seems that additional levels of regulation must determine the balance between somatostatin and GRF action. Among a variety of tissue-specific effects of somatostatin, it also acts on the pancreas to inhibit insulin and glucagon release.

SPEECH (OE): Greek **phanai** 'to speak' has given rise to several words in English: from the past participle **phatos** 'spoken': *aphasia*, the inability to speak, and *emphatic*; from **pheme** meaning 'speech': *euphemism, blasphemy,* and *prophet*. Several Latin words also have English relatives here: **dicere, dictum** 'to say' has given rise to *diction, edict, predict, contradict, dictionary*; **dictare, dictatum** 'to say with authority' has given *dictate* and *dictator*; **loquier, locutus** (L.) 'to speak, talk': *elocution, eloquent, circumlocution, colloquium, colloquy, loquacious, interlocutor, ventriloquist*; **fari, fatus** (L.)

'to say, predict': *preface, fable, fatal, fated*, also, from **in-** + **fans**, the present participle of **fari**: *infant*, young child or baby, literally 'unable to speak'.

SPORE: from **spora** (Gk.) 'sowing' or 'seed'. Spores are produced by bacteria, protozoa, algae, fungi, and plants. The function of spores is either to ensure *survival* under adverse conditions or to facilitate *dispersal*. Plant spores that function in seed dispersal are referred to as *diaspores*. *Diaspora* derives from the same root and refers to a human population scattered from its original homeland. Eukaryotic spores are usually haploid. Bacterial spores such as those of the *Anthrax* bacillus can survive in soil for decades.

Sperm similarly means seed both in animals and plants, as in *gymnosperm*, which describes plant seeds not enclosed in an ovary as opposed to *angiosperms*, which are. **Sperma** (Gk.) is 'that which is sown'.

SPORADIC *adj.*: **sporadikos** (Gk.) has the meaning 'scattered' in both Greek and English and derives from **speirein** 'to sow' and presumably is related to **dispersus** (L.), the root for dispersal. In a medical context, *sporadic* refers to a disease or complaint that occurs randomly and without predisposing circumstances and is contrasted with *familial*: 'Amyotrophic lateral sclerosis (ALS) is typically sporadic but around 10 per cent of cases are familial'. A sporadic disease may also be described as idiopathic, meaning it is unexplained or of unknown aetiology. *Idiopathic* has also been described as a useful synonym for 'nobody knows'. *Idio-* is a combining form from **idios** (Gk.) meaning 'own', or 'distinct', cf. *idiosyncrasy*.

See also SPOROZOITE under **APICOMPLEXA** heading.

STEM CELLS are distinguished from normal *somatic* cells by their ability to give rise to varieties of differentiated cells. *Embryonic stem* (ES) cells, derived from very early embryos at the eight-cell stage have the property of *pluripotency*, which is the ability to give rise to any of the more than 200 tissues of the human body. Other stem cells are located in particular tissues where their function is to differentiate into and maintain a limited number of tissues. Such *adult* stem cells are termed *multipotent*, and an example of these is the *haematopoietic* stem cell (HSC) system. See also HAEMATOPOIESIS entry under **BLOOD**. Apart from pluripotency, the other characteristic of stem cells is their relative *quiescence*, meaning that they do not often divide. Their immediate descendants when they do divide are referred to as *progenitor* cells. *Single-cell RNA sequencing* serves to identify cells produced in subsequent divisions in which particular RNAs first appear and thus foreshadow the order in which differentiation produces a series of distinct cell *lineages* (**linea** L. 'line of descent').

A further extraordinary discovery was that it is possible to reprogram mature somatic cells, such as skin *fibroblasts*, into pluripotent cells referred to as iPS cells or *induced pluripotent stem* cells. This was a development of work on embryonic stem cells where it was found that a set of twenty-four different *transcription factors* were involved in maintenance of ES pluripotency. Initial experiments with all twenty-four factors yielded occasional iPS cells. Eventually a procedure was found where just four factors were found to be sufficient.

Research on ES and iPS cells has revealed the ground rules for differentiation at the organ level. ES cells exposed to appropriate transcription and other factors can differentiate into neurons, immune cells, or beating heart cells. Human ES cells have been used to create a 'stomach in a dish'. The ES cells are actually grown out as *organoids* in 3D culture in response to a particular set of determinants of differentiation, such as transcription factors. These organoids contain all the cell lineages of mature gastric tissue: acid-secreting *parietal* cells, digestive-enzyme-secreting *chief* cells, *endocrine* cells, and *mucous* cells. (Parietal cells line body cavities or surfaces, from Latin **paries, pariet-** 'wall'.)

The timeline for these developments reflects the difficulties of the original work. 1981: successful culturing of mouse ES cells; 1995: ES cells from monkeys; 1998: using donated embryos that had gone unused from fertility treatments, the first human ES cell line; 2006: the first iPS cell line. Much of the early research on ES cells concentrated on making them easier to work with. They can now be produced quickly and reliably and show no sign of becoming cancerous. Clinical applications of ES cells are only just now being developed.[98]

STEP: gradus (L.) from the verb **gradi, gressus** 'pace, walk, or degree'; *grade, gradual, graduate, gradient, degrade* as well as *aggression, congress, digress, egress, progress, regress,* and *transgress.* **Bainein** (Gk.) 'to walk, go,

243

pass' gives *acrobat* in English from **acrobainein**, while its meaning in Greek was 'to walk on tiptoe'. Also, there is **diabainein**, literally 'to go through', giving *diabetes* in English, alluding to the symptoms of increased thirst and urination that characterize diabetes. As well as *acrobats*, we also now have *aerobatics*, and we could also have *hypnobats* for sleepwalkers and *hydrobats* for those insects that skate on the surface of water, while on the radio an excited young lady is singing 'I'm walking on sunshine'—maybe a *heliobat*.

STOCHASTIC: from the Greek, stochastic mechanisms or reactions are those that are randomly determined. Can fluctuation in metabolic activity due to stochastic effects lead to phenotypic heterogeneity? It appears that for individual cells of a bacterial culture, cellular metabolism is inherently stochastic. On the other hand, *homeostasis* is the rule in virtually all living systems, and this reflects the presence of stabilizing regulatory mechanisms that operate at the tissue and organism level and probably at the cell level as well in most situations. There remains the possibility that stochastic variation may enable cells in some situations to exploit new growth conditions.[99]

SULPHUR: **sulfur** (L.) but also **sulphur** and **sulpur** in Latin manuscripts: The traditional British spelling has been *sulphur*, apparently on the assumption that it derived from Greek because the letter *phi* (φ) transliterates as *ph*, an error evidently made by mediaeval scribes and possibly earlier than that. The International Union of Pure and Applied Chemistry (IUPAC) has now decreed that it is *sulfur* with an *f*, similarly sulfate, sulfite, sulfide, etc. For the time being, I will stick with sulphur. Sulphur the element has atomic number 16, electronic configuration 2,8,6 and is therefore below oxygen in the periodic table. The Greeks certainly knew about sulphur, but their word was **theion**, from which we get *thiol* as in thiol group (-SH), the sulphur equivalent of an -OH group, also *thiocyanate* (CNS), *thiosulphate* $S_2O_3^{2-}$ (some interesting etymology there), and *thiourea* $S=C(NH_2)_2$, the sulphur analogue of urea. Thiols are also referred to as *mercaptans* (L. **mercurium captans**), because they react readily with mercury salts, e.g. *mercaptoethanol* ($HO-CH_2-CH_2-SH$). Sulphur gets multiple mentions in the Bible as brimstone, usually in the context of sinners getting their comeuppance in the form of fire and brimstone. The amino acids *methionine* and *cysteine* are the two S-containing amino acids found in proteins, while oxidation of two cysteine -SH groups leads to the formation of a *disulphide bond*. *Thiamine*, the vitamin, is so named because it contains a sulphur atom. *Coenzyme A*, often abbreviated as CoASH, also has a free -SH group which upon reaction with an acyl group forms a *thioester* intermediate, e.g. acetyl CoA. See also **ANTIOXIDANTS** for the tripeptide *glutathione*.

Chromium, like sulphur and selenium, occurs in group VI of the periodic table and, like these, can exist in the +6 oxidation state, CrO_4^{2-}. Hexavalent chromium compounds have a bad reputation as carcinogens. Cells take up chromate using the sulphate transporter. Health hazards associated with the dumping of chromate waste were highlighted in the film *Erin Brockovich*.

SYMBIOSIS: from the Greek, where it meant 'living together'. In a biological context, it refers to interaction between two organisms. Variations depend on whether the interaction is permanent and whether it is to the advantage of one or both organisms. Familiar examples are mitochondria and chloroplasts, which are the descendants of bacteria and blue-green alga which invaded ancestral eukaryotic cells. See entries for **MITOCHONDRIA** and **PHOTOSYNTHESIS**. These are referred to as *endosymbionts* since they are permanently intracellular. Another example is the symbiotic association between nitrogen-fixing *Rhizobia* bacteria that occur within root nodules of leguminous plants.

Other examples of symbiosis can be described under the heading of *mutualism*, where the interaction benefits both parties as in the case of the gut *microbiome* that helps digestion. Another example is the association between a *fungus* and a *photosynthetic alga* that occurs in *lichens*, where the fungus supplies physical support and moisture. Recent genomic analyses have revealed that many lichens also contain yeast cells but whether these contribute to lichen growth is not clear. The other class of symbionts are the *parasites*, which include worms, protists such as malaria, bacteria, and viruses. **Parasitos** (Gk.) was 'a person eating at another's table'.

TERMINOLOGY: derives from **terminus** (L.) meaning 'end, limit, or boundary' but entered English from German, where it had acquired its current meaning. The phrase *terminological inexactitude* is said to have been first used in a speech by Winston Churchill in 1906 where he apparently meant what it says: inexact use of terminology. Subsequently it has become a *euphemism* for *bare-faced lie*: **eu-** 'well' + **pheme** 'speaking'; cf. *blasphemy*. **Terminus** has also entered English directly as *terminus* meaning 'boundary or end' as in 'end of the line' and as the 5'- or 3'- terminus of a polynucleotide or as the *N-terminal* and *C-terminal residues* of a polypeptide chain, as well as *computer terminal*. The corresponding verb **terminare, -avi, -atum** has given rise to English *terminate* and *termination* as in the termination or stop codons in mRNA, UAA, UAG, and UGA that signal the end of the message. From the same root are *determine, exterminate*.

TISSUE derives originally from the Latin **texere, texui, textum** meaning 'to weave'. *Text* and *texture* are from the same stem. *Tissue* is from the OF version of **texere**, i.e. **tistre**, and its past participle, **tissu**. Tissue had various meanings, originally a rich kind of woven material, later any kind of cloth, fabric, web, or network. The first *OED* reference to biological tissue is not till 1808. In higher animals, tissues are identified as glandular tissue, connective tissue (including cartilage, tooth, and bone), muscular tissue, and nervous tissue. Components of blood are sometimes referred to as fluid tissue.

TOMOS is the Greek root meaning 'a section', from the verb **temnein** 'to cut or divide' which in combination with **a-** meaning 'not' generated the English word *atom*, i.e. an indivisible entity, which was true at least in the days before particle physics. The original meaning of **tomos** was a book or a volume that was part of a larger work, hence the English word *tome*, but nowadays it usually indicates a large, heavy book. Most English words derived from **tomos** retain the original idea of 'to cut'. Examples are *anatomy*, an area of study that relies on body *dissection*, together with a large number of words that end in *-ectomy* 'to cut out', such as *appendectomy*, *tonsillectomy*, which refer to organs being cut out. A *microtome* is a piece of equipment used to cut very thin tissue sections for examination under a microscope. *Entomology*, the study of insects, comes from the Greek **entomon** meaning 'insect' and refers to the segmented body of insects, as does the word *insect* itself, which comes from the Latin **in** + **sectum,** the past participle of the verb **secare** meaning 'to cut or divide'. (Now is the time to ask yourself if you know the difference between entomology and etymology.) CAT scan stands for *computerized axial tomography*, and once again, the sense is of images obtained as a series of *sections*. Phlebotomy, **phleps, phleb-** (Gk.) 'vein' + **tomia** 'cutting', refers to the practice of taking blood, and the word has a somewhat mediaeval flavour from the days when the treatment for most complaints consisted of bleeding the patient, but doctors in British hospitals still send for the phlebotomist when ordering blood tests. Also, *dichotomy*: literally divided in two. However, *colostomy* derives from **kolon** (Gk.) 'large intestine' + **stoma**, 'mouth'. A final example is the word *epitome*, which comes from **epitemnein**, which had the meaning 'to abridge'. The meaning of the English word has wandered and refers to a person or thing that is the perfect example of a particular category. In the song by Gilbert and Sullivan in *The Pirates of Penzance* that begins 'I am the very model of a modern major general', G & S are saying he is the *epitome* of a modern major general. You could say that he is therefore a *cut* above the others.

TOPO-: the combining form derived from **topos** (Gk.) 'place', and the root for *utopia*. *Utopia* was the name of a book published in 1516 by Sir Thomas More.

It has been speculated that Sir Thomas was using a play on words. Since on the one hand, **ou-** 'not' + **topos** is not a real place, while on the other **eu** + **topos** is one where everything is good and the opposite of *dystopia*, then *utopia* is therefore the perfect place that doesn't exist. *Topographic* maps detail the physical features of a particular area, and *topology* is a branch of mathematics that deals with geometrical transformations of objects. Hence *supercoiled* circular DNA molecules that differ in the number of supercoils are referred to as *topoisomers*, and enzymes that induce supercoiling are *topoisomerases*. Details of the 3D arrangement of chromatin within the eukaryotic nucleus are referred to as *chromatin topology*. An *ectopic* pregnancy, **ek-** 'out' + **topos**, literally 'out of place', is a pregnancy that develops outside the uterus. A *toponym* is a place name, usually based on a particular topographical feature, e.g. Spring Ridge. An *atopic* reaction, from **atopos** 'out of place' is a hypersensitive allergic reaction which occurs in a part of the body not in contact with the allergen, such as atopic dermatitis, i.e. eczema.

ISOTOPES: **iso** 'same' + **topos**, in the sense of isotopes of an element occupy the same place in the periodic table. These are atoms of the same element which differ in mass because of different numbers of neutrons in the nucleus. Most elements consist of a mixture of isotopes. Isotopes may be radioactive, such as ^{14}C, or stable, such as ^{15}N.

-TROPH-: trophe (Gk.) 'food, nutrition'.

ATROPHY: **a-** 'without' + **trophe**: 'the wasting away of tissue'. Starvation is obviously one cause of tissue or muscle atrophy, but as the term is currently understood, it can also result from lack of exercise or be the effect of disease or just old age.

AUTOTROPH: **autos-** 'self' + **trophe**: any organism able to grow on simple nutrients providing carbon, nitrogen, sulphur, phosphorus, and salts; cf. HETEROTROPHS, which require complex organic nutrients, being unable to synthesize essential metabolites.

AUXOTROPH: **auxein** 'to increase' + **trophe**: an auxotroph is an organism (either naturally occurring or a mutant) that is unable to synthesize a particular compound that is required for its growth.

EUTROPHICATION: **eutrophia** 'adequate nutrition', a term used in the context of algal or phytoplankton blooms in lakes and rivers or coastal waters because of the presence of excess nitrogen and phosphorus compounds in run-off containing fertilizers or sewage. The undesirable effects include fish kills due to lack of oxygen, while algae may produce toxins.

DYSTROPHY: refers to wasting diseases, most frequently in connection with various muscular dystrophies or myopathies. The most severe form is Duchenne muscular dystrophy (DMD), the result of mutations in a gene encoding the protein *dystrophin*. This gene is the largest human gene, the *primary transcript* being 2,400 kilobases (kb) in length and takes sixteen hours to transcribe. Removal of the *introns* that separate some seventy-nine *exons* yields a *messenger RNA* 14 kb long that codes for a protein 3,685 amino acids in length. The dystrophin gene is located on the X chromosome, so most affected individuals are male and because the gene is so large, mutations are relatively frequent, amounting to 1 in 3,500 male births. The dystrophin protein amounts to but a tiny fraction of muscle cell protein but is part of a complex that links the muscle fibre to the cell membrane. Absence of dystrophin, an essential component of the architecture of the muscle cell, leads to cell death and muscle wastage.

HYPERTROPHY: enlargement of an organ or tissue due to increase in the size of its cells. A condition that can result in sudden death in otherwise healthy young people is familial *hypertrophic cardiac myopathy*, the result of a mutation in cardiac myosin.

TROPHIC LEVEL: indicates where an organism is in the food chain. The three basic ways that organisms obtain food are (a) as producers or *autotrophs*—plants and algae; (b) as consumers or *heterotrophs*—*herbivores*, *carnivores*, and *omnivores*; (c) as decomposers or *detritivores*—bacteria and fungi.

TROPHOBLAST: a layer of cells that surrounds the blastocyst and supplies nutrients to the early embryo. See **BLAST** entry.

TROPHOZOITE: the feeding stage of a parasite such as malaria.

TROPHY WIFE: the derivation here is not from ancient Greece but modern America and is defined as 'a young, attractive wife regarded as a status symbol for an older man'. Try to think of an example.

TROPICS, TROPISMS, and various other **TROPES,** from the Greek stem **tropein,** meaning 'to turn or affect', show up in English in many different contexts. The Greek word **tropikos** referred to the summer and winter solstices, corresponding to the longest and shortest days of the year when the sun 'turns'. The word *tropic* had arrived in English by Chaucer's time. The Tropic of Capricorn is the southernmost circle of latitude, where the sun can be directly overhead. The northern equivalent is the Tropic of Cancer. These are located approximately 23° south and north of the equator respectively. The regions between these two latitudes are the *tropics*.

TROPISMS in plants describe the curvature of plants in response to a particular stimulus such as light (*phototropism* or *heliotropism*), gravity (*geotropism*), or injury (*traumatropism*). These were studied as far back as the seventeenth century, but it was Charles Darwin and his son Francis who at the time carried out the most wide-ranging studies of plant tropisms and showed that the mechanism of curvature was differential growth. Later investigators found that light- and gravity-responsive material was transmitted to other parts of the plant and this eventually led to the discovery of plant hormones.

TROPISM, in a medical context where infectious diseases are concerned, denotes the preference of the infectious agent for a particular cell type. Hepatitis C is described as *hepatotropic*, meaning it infects the liver. The poliovirus is *neurotropic*. Similarly, HIV-1 and HIV-2 specifically infect cells expressing the CD4 surface antigen (T helper cells and macrophages). This specificity is termed *viral tropism*. Also from the recent literature: 'Mouse rotavirus has a specific tropism for the small intestine', and '*Plasmodium vivax* (malaria) shows a strict host tropism for reticulocytes'.

ALLOTROPY: **allos** 'other' + **tropos**: refers to the property of some chemical elements that exist in different structural forms. The best known examples are the allotropes of carbon; in one case, the atoms are bonded in a tetrahedral arrangement producing diamond, in the other as hexagonal sheets of graphite. Sulphur and phosphorus also occur as allotropes, as does oxygen in the form O_2 as well as O_3 ozone.

ATROPINE: a muscle relaxant and a poison obtained from the deadly nightshade plant *Atropa belladonna*. Atropine is used to dilate the pupil of the eye to facilitate its examination. The plant, in turn, gets its name from Atropos, literally 'without turn', one of the three goddesses of fate and destiny in Greek mythology. 'The lady's not for turning' is a well-known quote from a speech by Margaret Thatcher. Atropos, apparently, was even more formidable than Mrs T.[100]

AZEOTROPE: **a-** 'no', **zeo-**, 'boil', **tropein** 'to turn', meaning 'no effect of boiling'. An azeotrope consists of a mixture of two liquids that cannot be fractionated by distillation because the vapour has the same composition as the mixture. See under **ETHYL-**.

CHAOTROPIC agents are compounds that disrupt the folded structure of macromolecules. These include any compounds that interfere with the weak interactions that stabilize proteins, such as hydrogen bonds, hydrophobic interactions, and salt bridges. Concentrated solutions of urea or guanidine hydrochloride which disrupt hydrogen bonds are commonly used protein denaturants, likewise nonpolar solvents such as phenol that weaken hydrophobic interactions.

HYDROTROPY: **hydros** 'water' + **tropos**: describes the situation in which substances that are more or less hydrophobic and only slightly soluble in water dissolve readily in an aqueous solution containing a *hydrotrope*. Hydrotropes are *amphiphilic* molecules, and it has recently been shown that ATP, with its hydrophobic adenine ring and hydrophilic sugar + triphosphate, has a remarkable ability to prevent denatured proteins from aggregating.[101] Because the cellular concentration of ATP at around 5 mM is much greater than the micromolar concentrations required for metabolic reactions that utilize ATP, the suggestion is that ATP plays the additional role of maintaining protein solubility in the cytoplasm, where total protein concentration can be in excess of 100 mg/ml. The possible significance of these findings for neurodegenerative diseases that are characterized by amyloids consisting of protein aggregates was also noted.

PLEIOTROPY: **pleion** or **pleon** 'more' + **tropos**. This describes the situation where gene mutations can result in multiple phenotypic effects. The basis of the effect is that gene products, as well as drugs, can have alternate fates or functions in different tissues. The following quote is an example of the use of pleiotropy (but might be taken as an example of how not to use it): 'Although some drugs in clinical use are capable of augmenting autophagy, these compounds exert pleiotropic effects, revealing an unmet need to develop specific inducers of autophagy'. (What they are trying to say is 'Drugs currently used to stimulate autophagy have side effects'.)

SOMATOTROPE: refers to the cell lineage in the anterior pituitary that synthesizes *somatotropin*, better known as *growth hormone*. Other cells in the anterior pituitary that synthesize thyroid-stimulating hormone and prolactin etc. are *thyrotropes*, *lactotropes*, etc. There is some confusion in the earlier literature when it comes to the spellings of hormones such as *somatotropin* and *corticotropin* and the cells that produce them. The alternative spelling is *somatotrophin* and *somatotrophe*. The **trope** root means a turn or turning, while the **trophe** root means food or nourishment. The rule seems clear: use *somatotropin* for the hormone which 'turns' genes on, as well as for the cell that synthesizes the hormone, while confining use of the **trophe** root to words such as *auxotroph* that have to do with nutrition.

VIRUS: the term was borrowed from Latin, where it meant poison or venom. By the seventeenth century, viruses were recognized as agents of disease[58] and later were being referred to as *filterable viruses* or *ultramicroscopic viruses*. As recently as 1940, an article in the *Lancet* asserted there were two sorts of viruses, the ones you could see under a microscope, and the ones not visible under a microscope. In this the *Lancet* was following the dictum of Pasteur: 'Every virus is a microbe'. Invention of the electron microscope in 1945 settled the difference between viruses and bacteria.

Viruses are genetic elements that are only able to replicate within living cells. All varieties of prokaryotic and eukaryotic organisms are subject to viral infections. Viral genomes may consist of DNA or RNA, either single-stranded or double-stranded. Viral replication is *autonomous*, i.e. it is independent of host cell replication, and results in *high copy numbers* of the invading genome. This was exploited by modifying bacterial DNA viruses to facilitate DNA cloning and for early sequencing methods.

For single-stranded RNA viruses, the RNA may correspond to messenger RNA, termed + strand viruses; in others, it is the complement of the coding strand. In the case of influenza virus, the genome is *segmented*, consisting of eight separate non-coding strand RNAs. Most RNA viruses rely on an RNA-dependent RNA polymerase for replication of their genome; negative strand viruses necessarily contain this polymerase within the infecting virus particle. While replication of DNA is extremely accurate because of a *proofreading* mechanism that detects and corrects errors as part of the replication process, RNA-dependent polymerases have no proofreading mechanism, so RNA viruses accumulate mutations at a relatively rapid rate and this worries epidemiologists. The appearance of COVID-19 and flu variants are familiar examples and often results in vaccines being less effective. Other problems with the flu virus are that there are strains that infect other species such as birds and pigs, and its segmented genome means that if a cell happens to be co-infected by two different viruses then reassortment of the segments can occur with unpredictable consequences.

Another type of RNA virus is termed a *retrovirus*, which carries into the infected cell the enzyme *reverse transcriptase* which, as the name suggests, is able to convert its RNA genome into double-stranded DNA. Another virus-coded enzyme, *integrase*, may then insert at random the viral DNA into the host genome. The best studied retrovirus is HIV, the *human immunodeficiency virus*, responsible for AIDS. The term *viral tropism* refers to the type of cell that a particular virus can infect, and for HIV, these are immune cells including T-cells and macrophages. The HIV *provirus* may remain dormant until activation of the immune system occurs because of another infection, and it is these conditions that tend to activate replication of the provirus, leading to death of the immune host cell.

The size of viral genomes is extremely variable. The now-extinct smallpox genome consisted of 186,000 DNA base pairs, and the current pandemic corona virus has about 30,000 nucleotides of single-stranded RNA. Many viruses have small genomes that encode only a handful of genes. These include a couple of genes that enable infection by binding to cell surface molecules, a *polymerase*, genes involved in assembly of the virus, including a protein *capsid* that protects the genome. Animal viruses are often surrounded by a *lipid envelope* derived from the host cell *plasma membrane*. Viruses are ubiquitous and, in many cases, have influenced the evolution of their hosts. Recent techniques for identifying viruses reveal they are extraordinarily abundant. In the oceans, the estimate is that the total number of viral particles at any one time is 10^{31}, in terms of mass the equivalent of seventy-five million blue whales, and likely to have major effects on food webs, ecosystems, and even the atmosphere.

Beginning in 2003, a series of *megaviruses* has been isolated from amoeba. The biggest is nearly a micron long, is visible under the light microscope, and has a genome in excess of 2.5 million base pairs (2.5Mb), which codes for 2,556 protein sequences. Extensive characterization of these viruses indicates they fit all the normal criteria for viruses, while 90 per cent of the protein coding sequences have no homologues in the databases. This has led to the controversial suggestion that these large viruses are the remnants of a primitive cellular lineage different from *Eubacteria*, *Archaea*, and *Eukaryota*.[102]

─────

VITALISM has a long history. The basis for the difference between the living and the non-living must have been a discussion topic for our earliest ancestors. Philosophers and theologians have not hesitated to instruct the masses on the subject. In general, the vitalist viewpoint has been that while both living organisms and non-living objects contain mass, only the former is possessed of some *non-material principle* that accounts for the difference between the two categories. Given the then virtually complete absence of knowledge of how biological organisms work, we need not be too critical of vitalist speculations down through the ages. No one familiar with the current state of biological science now answers to the label vitalist; however, mention of the word *consciousness* is still sufficient to bring the occasional vitalist out of the woodwork. The synthesis of urea from ammonium cyanate by Wohler in 1828 is often cited as a key 'end of vitalism' event in that it had been claimed by some that it would never be possible to synthesize from inorganic precursors compounds that occurred in living organisms. There is evidence that Wohler and his chemical colleagues at the time regarded the significance of his synthesis of urea as a nice piece of organic synthesis rather than having implications for vitalism. In any case, it was soon followed by the synthesis of many other biological chemicals, and it seems that vitalism was not being taken seriously among organic chemists by the 1840s.

The case of Pasteur is interesting. In 1858, fresh from his triumphant demolition of the theory of spontaneous generation, he was studying fermentation, the production of alcohol from sugar by yeast. Although other eminent scientists were prepared to argue that *fermentation* was the result of catalysts acting within cells, Pasteur believed this was a process that could only occur as a result of 'vital action' within an intact cell. It was not till 1897 that Buchner showed that a cell-free extract of yeast was able to convert sugar to alcohol and carbon dioxide. If one had to nominate a single event that spelled the end of vitalism, then Buchner's experiment would have to be it. Venter's construction of an entire synthetic *Mycoplasma* chromosome and using it to convert one bacterial species into another is a spectacular demonstration that life is all about chemistry. See entry for **MYCO-**.

VITAMINS: vita (L.) 'life'. These are low-molecular-weight compounds which are essential for normal good health but which have to be supplied in the diet because humans are unable to synthesize them. These were originally named *vitamines* because the first to be isolated, vitamin B1 or *thiamine*, was known to contain an amino group, but it soon became clear that other vitamins did not. The concept of a vitamin grew out of observations by the British Royal Navy as far back as the nineteenth century that the provision of fruit in the diet prevented the occurrence of *scurvy* during long sea voyages. Some vitamins (A, D, E, and K) are *fat-soluble* and can be stored in the body. The others, members of the B complex and vitamin C, are *water-soluble* and an excess is rapidly excreted. Most vitamins serve as precursors of *coenzymes* or enzyme *prosthetic groups*. See entry for COFACTORS subheading under **ENZYMES** for this distinction.

Vitamin B1 in its active form as *thiamine pyrophosphate* serves as a coenzyme in reactions involving the transfer of two-carbon compounds (including the key metabolic enzyme *pyruvate dehydrogenase*) that converts *pyruvate* to *acetyl CoA*. *Riboflavin*, vitamin B2, which is bright yellow in colour, **flavus** (L.) 'yellow', is converted into the electron acceptors *FMN* and *FAD*. *Niacin* or *nicotinic acid*, vitamin B3, is the precursor for the electron acceptors NAD^+ and $NADP^+$. Nicotinic acid and *nicotine* derive their names from the tobacco plant *Nicotiana tabacum*, which was named after the sixteenth-century Frenchman Jean Nicot, who introduced tobacco to France in 1560. Niacin was introduced as an abbreviation of <u>ni</u>cotinic <u>ac</u>id and vitam<u>in</u> when B3 was introduced as a food supplement so that customers would not think their food was being contaminated with nicotine or tobacco. Vitamin B5 or *pantothenic acid*, **pantothen** (Gk.) 'everywhere', alluding to its widespread occurrence, is a precursor of *coenzyme A*. Vitamin B6, or pyridoxine, in its active form *pyridoxal phosphate*, is a coenzyme for many reactions involving amino acids such as *transamination* and *decarboxylation*. Its name indicates that it is a derivative of *pyridine* and contains an *aldehyde* group that can form a *Schiff base* with amino groups. Vitamin B7, or *biotin*, serves as a prosthetic group for enzymes that perform *carboxylation* reactions such as the synthesis of malonyl CoA from acetyl CoA, the committed step in *fatty acid synthesis*. Vitamin B12 is known to be required for only two reactions in mammals. One is a step in the breakdown of odd-numbered fatty acids. The other is the regeneration of *methionine* from *homocysteine* following transfer of the methionine methyl group in *methylation* reactions. Absence of this reaction interferes with reactions including *purine* and *thymine* synthesis that are required for DNA synthesis. Vitamin B12 is constructed along the lines of *haem* and *chlorophyll*: a ring of four modified *pyrroles* coordinated to a central metal atom, in this case a *cobalt ion*. B12 is also referred to as *cobalamin*. *Folic acid*, named for its occurrence in plants, **folium** (L.) 'leaf', serves as a coenzyme in reactions involving the transfer of *one-carbon* moieties. It is sometimes called vitamin B9 but is not very soluble in water. Folic acid deficiency has been implicated in the occurrence of birth defects, and many countries now supplement flour with folic acid.

Vitamin A is *retinol* which is converted to *retinal* to form the light-absorbing group, i.e. the *chromophore*, of rhodopsin. See entry for **OPSINS** and for **ISOPRENE DERIVATIVES**. See the entry for **ASCORBIC ACID** for vitamin C and that for vitamin D under CHOLESTEROL DERIVATIVES. Vitamin E consists of a number of related *tocopherols* (**tokos** Gk. 'offspring' + **pherein** 'to bear', perhaps best translated as 'life-giving') that are antioxidants and prevent the reaction of reactive oxygen species (ROS) with *polyunsaturated acids*. See BLOOD CLOTTING and OSTEOCALCIN for the function of vitamin K.

Vitamin D, which is required for *calcium absorption* in the intestine, is another derivative of cholesterol. It is formed in the skin by the action of the long UV or UVB fraction of sunlight on cholesterol, which results in cleavage of the steroid B ring. The product *cholecalciferol,* termed a *sec-steroid*, undergoes hydroxylations in the liver and kidneys to form the biologically active species *calcitriol*. As described under the entry for NUCLEAR RECEPTORS, the binding of vitamin D to its receptor creates a *transcription factor* that enables a calcium *transporter* gene to be expressed in the intestine. Since humans are able to synthesize calcitriol, vitamin D is really a hormone rather than a vitamin.

WORDS of BECOMING (based on verbs known as INCEPTIVES). These are words that describe processes that occur gradually, over a period of time. Latin has hundreds if not thousands of words that are used to describe gradual processes, and these are paired with words describing the final state. For example, **frigere** means 'to be cold' and **frigescere** means 'to grow cold'. Latin verbs of this type all have the same *-escere* suffix. Some have come across into English with the addition of an appropriate suffix: *-esce* for verbs, *-escent* for adjectives, and *-escence* for nouns. The more common examples include the following:

ACQUIESCE meaning to agree or accept: from **ad-** + **quiescere** 'to rest' from **quies** 'quiet', similarly QUIESCENT meaning inactive or dormant. In the area of cell biology, *quiescence* is defined as the state of a non-dividing cell that is still metabolically active and able to return to the cell cycle with full viability.

ADOLESCENT: literally becoming an adult, from **adolescere** 'to grow up'.

COALESCE: from **coalescere** 'to unite' from **co-** + **alescere** 'to grow up. The past participle of **coalescere** is **coalitum**, from which we get *coalition*. **Alescere** in turn derives from **alere** 'to nourish', from which English has obtained *alimentary*, i.e. relating to *nourishment*, in particular the *alimentary canal*.

CONVALESCE, meaning to recover, get better: from **convalescere** 'to become strong' and **valere** 'to be strong', which also gave rise to *ambivalent, equivalent, prevail, prevalent, valency, valiant, valour, value*.

DELIQUESCE meaning to become liquid, from **deliquescere** 'to dissolve'. *Deliquescent*, the adjective, is used in chemistry to describe materials that absorb moisture from the air and in some cases may actually dissolve in it.

EFFLORESCE from **efflorescere** 'to flower or bloom', a meaning that has carried over to English. In Latin, the meaning was usually metaphorical, referring to youth or beauty. Efflorescence has acquired a second meaning, referring in chemistry to the phenomenon in which crystalline compounds containing water of crystallization lose water in dry air and the crystals convert to a powder. Crystals of sodium thiosulphate pentahydrate effloresce in dry air and deliquesce in moist air. **Florescere**, 'to begin to flower', also acquired a secondary meaning, 'to begin to prosper' or 'flourish', the latter derived via OF.

EFFERVESCE meaning to foam, froth, boil: from **ex-** + **fervescere** 'begin to boil' from **fervere** 'to be hot, boil', which also gave rise to *fervent, fervid,* and *fervour*. The *fer* stem also shows up in *ferment* and *fermentation*. The Latin for yeast was *fermentum*. See entry for **YEAST**.

EVANESCE to fade or vanish, from **evanescere** 'to fade or pass away', thus evanescent *adj.* meaning ephemeral and evanescence *n.* as in 'the evanescence of fame'.

EXCRESCENCE: an abnormal outgrowth on an animal body or plant or any abnormal increase, or metaphorically as in *the excrescence of inequality*, from **excrescere** 'grow out' from **crescere** to grow. **Crescere** has given rise to *crescent*, as in *crescent moon*, describing the shape of the moon as it waxes and wanes. Also derived from variants of **crescere** are *increase, decrease, increment*. **Recrescere** 'to grow again' is the origin of *recruit*, via OF.

INCANDESCENT meaning to be hot, or hot enough to emit light or glow, from **incandescere**, also **candescere**, both meaning to become (white) hot.

OBSOLESCENT meaning to become obsolete, i.e. out of date or superseded, from **obsolescere** 'to become obsolete' and possibly related to **olere** 'to smell'.

PUBESCENT: reaching the age of puberty, from **pubescere**.

PUTRESCENT 'to go rotten or putrid', from **putrescere**.

RECRUDESCENT 'relapsing after treatment for a disease', from **recrudescere**.

SENESCENT from **senescere** to grow old. The study of **SENESCENCE** is a current hot research topic—and likely to remain so for the foreseeable.

TUMESCENT 'beginning to swell', from **tumescere**, hence *tumour.*

In the same manner, some relatively new useful technical words, mainly to do with light, have been invented:

FLUORESCENCE: is the property displayed by some materials of emitting radiation, either in the visible or invisible range, as a result of absorption of radiation of shorter wavelength. Fluorescence is a kind of luminescence but differs from phosphorescence in that it does not persist once the irradiating shorter wavelength light is turned off. The word was coined by Stokes in 1852 after observing fluorescence of fluorspar.

IRIDESCENT: displaying colours like those of the rainbow, from **iris** 'rainbow' in both Latin and Greek. The colours tend to change when viewed from different angles. The element iridium is so named on account of the iridescence it displays when dissolved in hydrochloric acid.

LUMINESCENT: having the property of emitting light, otherwise than as a result of incandescence. First recorded in the *OED* in 1889, derived from **lumen, lumin-** (L.) 'light'. Bioluminescence is the emission of light by organisms such as fireflies and many fish and is also an example of chemiluminescence.

OPALESCENT: exhibiting various colours like those of an opal. First use: 1813 (*OED*).

PHOSPHORESCENT: having the property of shining in the dark, derived from the behaviour of white phosphorus, P_4, which glows in the dark. The earliest use of phosphorescent, according to the *OED*, was 1766.

There are numerous common words of similar derivation but which have lost the **-escere** tag, e.g. *accrue* from **accrescere** 'become larger'. Also *increase*: via OF from **increscere** from **in-** + **crescere** 'grow'; also *increment* and *crescendo*. See EXCRESCENCE above.

There must be lots of other words of becoming that need inventing. Here are some suggestions with possible meanings:
Expensivescent: house prices.
Inebriescent: it won't be long now before the fight starts.
Delinquescence: kids!
Activescence: it always takes me a while to get going in the morning.

XENO-: xenos (Gk.) 'stranger or foreigner'. AXENIC: **a-** 'expressing absence' + **xenikos** 'of a stranger or alien'. Axenic cultures are ones that are free of contaminating organisms. They are usually obtained from mixed cultures by a series of dilutions or, in the case of bacteria, selecting single colonies from an agar plate.

XENARTHRA: **xenos** + **arthron** 'joint'. Order of mammals which includes anteaters, armadillos, and sloths, all from Central or South America, that lack canine and incisor teeth. Xenarthra means 'strange joint' and refers to their vertebral joints which have extra articulations and are unlike those of any other mammal. In an analysis of the evolutionary history of the placental mammals, it was found that the initial split among the placentals involved the *Xenarthra* and *Epitheria*, the latter being defined as all other placentals.

XENOBIOTIC: a word fashioned after the style of antibiotics: xenobiotics are compounds that are not normally found within a particular organism. The two main types of xenobiotics are drugs, medicinal or otherwise, and chemical pollutants that contaminate air, water, and food. See the entry for P450 CYTOCHROMES under **CYTO-** for how higher organisms deal with xenobiotics.

XENOGRAFT or XENOTRANSPLANTATION: a tissue or organ transplant from a species different from that of the recipient. At one time, it was suggested that kidneys from pigs could be transplanted into humans, but this line of investigation was largely abandoned because of the possibility that this could result in the activation of porcine RETROVIRUSES, as well as the likelihood of immune rejection. The development of new *immunosuppressants* and the ability to inactivate retroviruses using techniques such as CRISPR has led to renewed interest in this field, with human trials possible in the near future.[103] ALLOGRAFT refers to transplantation between two members of the same species that are not genetically identical, while HOMOGRAFT is where they *are* genetically identical. The transmissible facial tumour that afflicts the Tasmanian devil is an example of an allograft. This was shown by examining the karyotypes of eleven different tumours, all of which were identical and markedly different from the normal host chromosome pattern.

XENON: chemical element number 54, one of the *inert gases* and one of the rarest elements on earth. It was discovered in 1898 by the fractional distillation of liquid air by Ramsay and Travers. They called it xenon on account of its elusiveness. They had already discovered, by the same method, the other inert gases to which they also gave appropriate names: neon (**neos** Gk. 'new'), argon (**a-** 'without' + **ergon** 'work', named on account of its lack of reactivity), and krypton (**krupton** 'hidden').

XENOPHAGY: see heading for AUTOPHAGY under **PHAGE**.

XERO- **xeros** (Gk.) 'dry'. XERODERMA PIGMENTOSA (XP): **xeroderma** 'dry skin' + **pigmentosa** (L.) 'pigmented'. XP is an autosomal recessive disease caused by defects in any one of several different genes that specify proteins involved in the repair of DNA damage caused by UV light. XEROGRAPHY: **xeros** + **graphos** 'writing', literally 'dry writing', i.e. photocopying. XEROPHYTE: **xeros** + **phyton** 'plant', a plant adapted to growing in arid conditions and able to survive on minimal amounts of water.

XYLO: **xulon** (Gk.) 'wood'. XYLEM: vascular tissue in plants that conducts water and nutrients from the roots. XYLOSE: a D-pentose sugar occurring mainly in *hemicelluloses* that represent 20 to 30 per cent of wood, depending on the type of tree. Another pentose is *lyxose*, which is a C2 *epimer* of xylose, and the names for these two sugars are obvious *anagrams*. Two other pentoses are *arabinose*, first isolated from gum arabic, and its C2 epimer, *ribose*. It is said that the name *ribose* was arrived at by a partial rearrangement of the letters in arabinose. If ribose had been discovered first and the anagram rule applied, then arabinose would have been called birose, which would have been an appropriate name for a *pen*tose. On the other hand, an anagram of arabinose is abrainose—a no-brainer. Deoxyabrainonucleic acid does not exactly trip off the tongue. The derivation of ribose from arabinose is at odds with the claim that it is derived from the word *rib* as suggested by the story in the Bible that God created Eve from one of Adam's ribs—'And the Lord accosted Adam . . .' (**costa** L. 'rib'). One school of biblical exegesis (Greek **exegesis** 'interpretation, explanation') asserts that the word *rib* is an abbreviation of ribonucleic acid and that this is the first literature reference to the use of reverse transcriptase, the enzyme that synthesizes DNA using an RNA template.

Naming related compounds, particularly isomers, with names that are anagrams occurred occasionally in the early days of organic chemistry before systematic nomenclature was established, and the xylose/lyxose pair must be one of the few remaining examples. This practice was more widespread in botany. Stearn (1992) lists numerous examples as well as the protest of the Rev. R. T. Lowe in 1868 against 'such scandalously childish, bald and witless trickery with names . . . such base name-coinage'. Perhaps he was known as Rev. Trowel.

YEAST: OE, of Germanic origin, but **zyme** in Greek, derived from an Indo-European root shared by Greek **zein** 'to boil'. Yeast is a single-cell *fungus* which has in recent times been intensively studied, serving as a model eukaryotic organism. One of the approximately 1,500 naturally occurring strains of yeast that has been most intensively investigated is *Saccharomyces cerevisiae*, literally 'sugar fungus for making bread'. Yeast has been used since ancient times in wine and beer manufacture as well as in bread making. During *alcoholic fermentation*, yeast converts the six carbons in a molecule of sugar into two molecules of ethanol and two molecules of carbon dioxide, accompanied by effervescence and the generation of heat, the word for yeast in Latin being **fermentum**. During bread making, it is the generation of carbon dioxide that causes the dough to rise. Until the eighteenth century, not yeast but *leaven*, from **levare** 'to raise', was the word used to describe the agent responsible for these processes. A more familiar word is *unleavened*, describing flat bread that has been made without yeast. The first citation for *yeast* in the *OED* is 1743 from the *London and Country Brewer*: 'Yeast consists of a great quantity of subtle and spirituous particles, wrapped up in such as are viscid'. In those days, the action of leaven was the result of a mysterious influence similar to that attributed to the agent of an influenza epidemic—see entry for INFLUENZA under **FLUX**. Thus *OED* 1758: 'Her blood was loaded with a bad leven'. From the *OED* 1876: 'The ancients used as leaven for their bread either dough that had been kept till it was sour, or beer yeast', hence the origin of the term *sourdough*.

ZOONOSES: **zoo-** (Gk.) 'relating to animals' + **nosos** 'disease': diseases transmissible to humans from animals are termed *zoonotic* diseases. Many human diseases are in this category, including recurring flu epidemics that have originated from birds or pigs. Most types of animals can serve as reservoirs of human disease, and transmission to humans is most likely where animals live in close proximity. According to the Australian biologist Tim Flannery, 'we share twenty-six diseases with poultry, thirty-two with rats and mice, thirty-five with horses, forty-two with pigs, forty-six with sheep and goats, fifty with cattle, and sixty-five with our oldest companion, the dog'.

Bats, although not a source of day-to-day contact, are also a source of viruses that infect humans, including Ebola and various coronaviruses, in particular the SARS COVID-2 virus responsible for the COVID-19 *pandemic* that began in China in 2019 (SARS: *severe acute respiratory syndrome*). Fortunately, COVID-19 is susceptible to vaccination, although in many parts of the world the virus arrives well before supplies of vaccine. The appearance of the more infectious variants compounds this problem and highlights the possibility that new variants resistant to current vaccines may arise.

Some survivors of COVID-19 infection continue to suffer after-effects for a prolonged period, known as long COVID. A possible reason for this has to do with a long-term effect of viral infection on the immune system. In the case of infection by COVID-19 involving a high virus load, there can be so much tissue injury that the immune system cannot figure out initially whether it should be recognizing the virus or the self-antigens being released from infected cells.[104] See note 17 for more on this as well as under **AUTOIMMUNITY**.

But COVID-19 is only the tip of the iceberg. It is estimated that around 89 per cent of the 180 recognized RNA viruses with the potential to harm humans are zoonotic.[105] See also under **VIRUSES**.

NOTES

1. Kandel (2006), 102. The line that biochemistry provides a universal language of biology, which I have adopted in the title of this book, comes from the autobiography of the Nobel Prize winner Eric Kandel, *In Search of Memory: The Emergence of a New Science of Mind*. He was impressed by the current advances in neuroscience showing that the workings of the brain involved both electrical and chemical signals: 'In addition, since biochemistry is a universal language of biology, synaptic transmission piqued the interest of the biological community as a whole, not to mention students of behavior and mind, like me'.
2. Bragg (2003), 21. Material in this chapter is also covered in Bryson (1991).
3. Bragg, 38.
4. Ibid., 7.
5. Leonhardt (2013), chapter 3.
6. Montgomery (2013), 135 et seq.
7. Leonhardt, 136–142.
8. McLeish, T. C. B. (2014). 'A Medieval Universe', *Nature* **507**, 161–163.
9. Alexander, C. (2009), xiii
10. Kerr, J. F. R. (1972) 'Apoptosis a basic biological phenomenon', *Brit. Jrnl. Cancer* **26**, 239.
11. Bynum, W. F. (2013), *Nature* **495** 169: on the career of John Snow (1813–1858), who pioneered the use of ether and chloroform as anaesthetics in London.
12. Burnside (2009), 238.
13. Michael Faraday was the first to isolate *benzene* in 1825 from the oily residue obtained from the production of gas for lighting. His name for it was bicarburet of hydrogen because he thought it contained twice as much carbon as hydrogen. In 1833, the French chemist Laurent introduced the name *phene* as the name for benzene on account of its occurrence as a by-product of gaslight production, the rationale being its derivation from *phanein* (Gk.) meaning 'to cause to be seen'. Subsequently, phene gave rise to phenyl-, which is the radical obtained by extraction of a hydrogen atom from benzene, and so to the amino acid *phenylalanine* and *phenol* C_6H_5OH, and many other phenyl derivatives. A new crystalline allotrope of carbon, first isolated in 2004, is *graphene*, which consists of a two-dimensional array of benzene rings with remarkable properties, also described as a one-atom-thick layer of *graphite*.
14. The Anthropocene made it into the *OED* in 2014, defined as follows: 'the era of geological time during which human activity is considered to be the

dominant influence on the environment, climate and ecology of the earth'. The *OED* was soon in trouble for calling it an era—it should be an epoch. A final decision on how to define the Anthropocene is up to the geologists. It could be classified as the *epoch* that follows the Holocene (**holo-** 'whole' + **kainos** 'new') that began about 11,700 years ago, part of the Quaternary *period* that began about 2.5 million years ago within the Cenozoic *era* (**kainos** + **zoon** 'animal') that saw the evolution and dominance of mammals and began about 65 million years ago, lasting through to the present day and part of the Phanerozoic *eon* (**phaneros** 'visible'), which began around 540 million years ago with the evolution of multicellular organisms.

15. Global warming due to increased atmospheric carbon dioxide was predicted by the Swedish physical chemist Svante Arrhenius in 1896. He was not overly concerned by this because he thought it would take 3,000 years of coal burning to increase carbon dioxide levels by 50 per cent. Instead, the level of carbon dioxide increased by 30 per cent during the twentieth century.

16. Spang et al. (2015). 'Complex archaea that bridge the gap between prokaryotes and eukaryotes', *Nature* **521**, 173–179. Based on genomic sequencing, several archaeal lineages have been identified as likely intermediates in the emergence of eukaryotes. These have been lumped together as *Asgard archaea*. A recent paper reports that it has now been possible to culture one such species: Imachi H et al. (2020), 'Isolation of an archeon at the prokaryote-eukaryote interface', *Nature* **577**, 478–479 and 519–525.

17. See 'Outlook' at *Nature* **595**, No. 7867 Suppl., S45–S66 (2021) for summaries of recent investigations into autoimmunity.

18. According to Barnes (2002), 89, the use of erythropoietin around the year 2000 by competitors in the Tour de France led to what was christened 'the two-speed *peloton*'—those using it and those not.

19. Garner, H. and de Visser, K. E. (2017). 'Neutrophils take a round trip', *Science* **358**, 42–43. These authors discuss the possible role of neutrophils in chronic inflammation.

20. Libby, P. (2021). 'The changing landscape of atherosclerosis', *Nature* **592**, 524–533.

21. Mattern (2013), 246.

22. Kalinich, M. and Haber, D. A. (2018), 'Cancer detection: seeking signals in blood', *Science* **359**, 866–7; Ma, N. and Jeffrey, S. S. (2020), 'Deciphering cancer clues from blood', *Science* **367**, 1424–5.

23. Li, L. et al. (2019). 'p53 regulation of ammonia metabolism through urea cycle controls polyamine biosynthesis', *Nature* **567**, 253–256. Polyamines are a group of small basic compounds including cadaverine, putrescine, spermidine, and spermine. Being polycations, they bind strongly to negatively charged DNA and RNA and are abundant in rapidly proliferating cells. Polyamine

synthesis occurs in the cytoplasm of all cells, and if polyamine synthesis is inhibited, cell growth stops or is severely retarded. Polyamines have been implicated in a range of processes including transcription, translation, cell signalling, and modulation of chromatin structure—perhaps by boldly going where no histone has gone before.

24. Several papers on the 2,658 genomes are contained in *Nature* **578**, 89–136 and summarized on pp. 39–40, issue of 6 February 2020.

25. Wapner (2013). This book *The Philadelphia Chromosome*, as well as providing an account of the development of Gleevec and its success in treating chronic myeloid leukaemia, also describes the early work that led to the concept of the oncogene.

26. Sugar solutions are invariably colourless and their crystals white. It is not clear why erythrose is so named, **eruthro** 'red' as in erythrocytes. One suggestion is that it was first isolated from rhubarb.

27. Parker, H. G. et al. (2009), 'An Expressed Fgf4 Retrogene is Associated with Breed-Defining Chondrodysplasia in Domestic Dogs', *Science* **325**, 995–998. Retrotranscription involves the copying by the enzyme reverse transcriptase of a messenger RNA molecule into DNA which is then inserted back into the genome. Transcription of the resulting *retrogene* is now under the control of a different promoter accounting for overexpression of Fgf4. See also entry for RETROTRANSPOSONS.

28. Ball, P. (2001), 147.

29. Crosland (1962), xv.

30. Marshall, M. (2021). 'The Pace of Development', *Nature* **592**, 682–684.

31. Several reviews on circadian biology are contained in *Science* **354**, 987–1015 (2016), issue of 25 November 2016. Also, Stokstad, E. and Vogel, G. (2017), 'Revelations about rhythm of life rewarded', *Science* **358**, 18. A recent paper makes the case for increased attention to harmonizing drug dosing times with the circadian clock. Experiments in animals indicate this is important, but more human trials are needed. Mice do not run on the same clock as humans. Ruben MD *et al.* (2019), 'Dosing time matters', *Science* **365**, 547–549.

32. Ball (2002). Ball also comments that most people would have difficulty distinguishing indigo and violet in the standard spectrum. It seems that Newton, in fixing on seven different colours, was influenced by the traditional idea that seven is a magic number. If in doubt about that, search Google for 'the number seven'.

33. Kupferschmidt, K. (2019), 'In search of blue', *Science* **364**, 424–429.

34. Milne, J. L. S. et al. (2013), 'Cryo-electron microscopy—a primer for the non-microscopist', *FEBS J.* **280**, 28–45. Further advances in cryo-EM are summarized in Herzik, M. A. (2020), 'Cryo-electron microscopy reaches atomic resolution', *Nature* **587**, 39–40.

---

35. Vermeij, W. P. and Hoeijmakers, J. H. J. (2021), 'Base editor repairs gene of premature-ageing disease', *Nature* **589**, 522–524.

36. Bowman (2003). This book describes the discovery at Vindolanda, near Hadrian's Wall, of a large collection of written material preserved on wooden tablets from the time of the Roman occupation. The quote on p. 23 is as follows: 'The contemptuous term *Brittunculi*, which is a recognizable diminutive formation . . . suggests no great sympathy for the subjects of Roman rule'. One is reminded of the old joke about Julius Caesar, who after a victorious campaign declared '**veni, vidi, vici**' (I came, I saw, I conquered). It has been claimed that he was describing the inhabitants of Britannia as 'weeny, weedy, and weaky'.

37. The apparent ambiguity surrounding both the Latin and Greek words for speaking and reading may have something to do with an ongoing argument about whether or not it was usual for the ancients to read aloud. Some evidence that reading silently was not routine comes from the comment made by St Augustine when he met St Ambrose, the bishop of Milan in 383 CE. 'When he read, his eyes scanned the page and his heart explored the meaning, but his voice was silent and his tongue was still.' Saint Augustine, p. 114.

38. Cited by Crosland, p. 62, from p. 187 of the 1911 Everyman edition of *The Sceptical Chymist*.

39. Kruger, K. et al. (1982), 'Self-splicing RNA: Autoexcision and autocyclization of the ribosomal RNA intervening sequence of *Tetrahymena*.' *Cell* **31**, 147–157. Guerrier-Takada, C. et al. (1983). 'The RNA moiety of ribonuclease P is the catalytic subunit of the enzyme', *Cell* **35**, 849–857.
    Thomas Cech and Sidney Altman were jointly awarded the 1989 Nobel Prize in Chemistry for the work described in the above papers.
    Yarus, M. and Welch, M. (2000), 'Peptidyl transferase: ancient and exiguous', *Chemistry and Biology* **7**, R187–R190. Also, Yarus (2010).

40. Quoted by Dixon in Needham (1971).

41. Groupings such as sulphate had previously been recognized in inorganic chemistry and described as radicals by de Morveau in 1786. Liebig and Wohler, on the basis of similar observations on the benzoyl radical, introduced the term into organic chemistry in 1832. Crosland (1962), 302.

42. The term *zeolite* covers a large group of compounds both natural and synthetic. It was originally introduced when it was observed that heating the aluminosilicate *stilbite* produced steam from adsorbed water. Derivation: **zeo** (Gk.) 'boil' + **lithos** 'stone'.

43. Kohl, K.D. et al. (2014). 'Gut microbes of mammalian herbivores facilitate intake of plant toxins', *Ecol. Lett.* **17**, 1238–1246.

44. The extent to which the reversible reaction A + B → C + D proceeds under a

particular set of conditions is given by the ratio of products to reactants, from which it follows that the free energy of the reaction ΔG is defined as $\Delta G = \Delta G^{0\prime} + RT \log_e [C][D] / [A][B]$, where $\Delta G^{0\prime}$ is a *standard free energy change* under defined conditions of pH and temperature, R is the gas constant, and T is the absolute temperature. At equilibrium, $\Delta G = 0$ and since $[C][D] / [A][B] = K'_{eq}$ then $-\Delta G^{0\prime} = RT \log_e K'_{eq}$ where K'_{eq} is the *equilibrium constant* under the defined conditions.

45. According to Wikipedia, the word *genome* was coined in 1920 by the botanist Hans Winkler by combining 'gen' with the 'ome' of chromosome. Apparently the -*ome* suffix was rampant in the botanical literature at that time, as judged by the following: *phyllome*: an assemblage of leaves; *hadrome*: the xylem or woody portion of a vascular bundle, consisting of the *hydrome* and part of the *amylome*; together with the *leptome*, it forms the *mestome* (Jackson 1900).

46. International Human Genome Sequencing Consortium (2001), 'Initial sequencing and analysis of the human genome', *Nature* **409**, 860–921. Further sequencing and analysis are summarized in Lander, E. S. (2011), 'Initial impact of the sequencing of the human genome', *Nature* **470**, 187–197. Several papers in *Nature*, issue of 11 February 2021, mark the twentieth anniversary of publication of the human genome sequence. That original version is now described as 'gappy and error-filled'.

47. This apparently happened in Japan about 100 generations ago, leading to the inactivation of the *fukutin* gene, causing mental retardation and muscle wasting in homozygotes. About one in ninety Japanese carry one copy of this retrotransposon insertion which is the cause of one of the most common diseases in Japan, see *Nature* **478** (2011), 46 and 127.
Gorbunova, V. et al. (2021), 'The role of retrotransposable elements in ageing and age-associated diseases', *Nature* **596**, 43–53.

48. De Koning, A. P. et al. (2011), 'Repetitive elements may comprise over two thirds of the human genome', *PLoS Genet.* **7**, e1002384.

49. Publications by the 1000 Genomes Project Consortium (2015): see *Nature* **526**, 68–74 and 75–81. For a summary of a more recent comprehensive study of human genetic variation, see Church, D. M. (2020), 'Ability to understand genomes scales up', *Nature* **581**, 385–6.

50. Dehal, P. and Boore, J. L. (2005), 'Two rounds of whole genome duplication in the ancestral vertebrate', *PLoS* **3**, e314.

51. Berthelot, C. et al. (2014), 'The rainbow trout genome provides novel insights into evolution after whole-genome duplication in vertebrates', *Nature Comm.* **5**, 1–10, article number 3657.

52. Nielsen, R. et al. (2017), 'Tracing the peopling of the world through genomics', *Nature* **541**, 302–310. Malaspinas, A. S. et al. (2016), 'A genomic history of Australia', *Nature* **538**, 207–214. Also, recent reviews: Bergstrom, A. et al.

(2021), 'Origins of modern human ancestry', *Nature* **590** 229–237. Willerslev, E. and Meltzer, D. J. (2021), 'Peopling of the Americas as inferred from ancient genomics', *Nature* **594**, 356–364.

53. The landmark paper for GWAS studies is the following: 'Genome-wide association study of 14,000 cases of seven common diseases and 3,000 shared controls', The Wellcome Trust Case Control Consortium (2007), *Nature* **447**, 661–678.

54. Heritability is a term used to describe the strength of inheritance of a character, i.e. whether it is likely to be passed on to the next generation. For example, height is strongly heritable, and several hundred genome sites have been shown to contribute to its heritability.

55. Bycroft, C. et al. (2018), 'The UK Biobank resource with deep phenotyping and genomic data', *Nature* **562**, 203–209 (open access). Wang, Q. et al. (2021), 'Rare variant contribution to human disease in 281,104 UK Biobank exomes', *Nature* **597**, 527–532.

56. *Nature* editorial (2018) **563**, 155–6.

57. Jinek, M. et al. (2013), 'RNA-programmed genome editing in human cells', *eLife* **7**, e00471. Cohen, J. (2019), 'China's CRISPR Revolution', *Science* **365**, 420–421, 422–425, 426–429, and 430–437. This last paper, p. 430, relates the circumstances of a very controversial gene-editing experiment carried out on human embryos in China.

58. In 1996, the 200th anniversary of Jenner's introduction of vaccination, details surrounding its further dissemination were published: Cook, G. C. (1996), 'William Woodville and vaccination' *Nature* **381**, 18 and Empson, J. (1996), 'Country doctor and speckled monster', *Nature* **381**, 26. It turns out that Jenner was already well known and had been elected fellow of the Royal Society for his paper 'Observations on the Natural History of the Cuckoo'. His subsequent manuscript on cowpox was rejected for publication by Sir Joseph Banks, the president of the Royal Society. Jenner published it in expanded form as a pamphlet. Woodville, superintendent of St Pancras Smallpox Hospital in London, organized the first large-scale clinical trial in London in 1799. He was later largely responsible for introducing vaccination to France. An international campaign for smallpox vaccination led, after nearly 200 years, to the announcement in 1980 by the World Health Organization of the global eradication of smallpox, the virus that had been a public health scourge, accounting in many situations for 10 per cent of total deaths and a third of those of children.

59. Each L chain contains two examples of a structure known as the *immunoglobulin* or *Ig fold* and each H chain four such structures. A protein fold is a particular arrangement of a polypeptide chain in three dimensions. The actual amino acid sequences in different examples of the same fold

may be quite variable but nonetheless sufficiently similar so that the overall 3D structure is conserved. Determination of the human genome sequence revealed that there are some 750 genes that encode proteins with at least one copy of the Ig fold. Many, but not all, of these proteins function in the immune system. The Ig fold and numerous other examples of protein folds are like Lego pieces—they can be used in many different situations.

60. *Somatic* cells are any cells except for the reproductive cells, from **soma** (Gk.) 'body'. The point here is that the rearrangements that generate antibodies, while inherited by descendants of the cell in which they occur, are not inherited by descendants of the organism.

61. Davis (2014), 94.

62. Davis (2014), 98. Davis (2018), 149.

63. Davis (2018), 184 et seq.

64. Small peptides containing six or eight cysteines are not uncommon in snake and spider venoms. See entry for cysteine under **AMINO ACIDS**.

65. Dolgin, E. (2021), 'The Tangled History of mRNA Vaccines', *Nature* **597**, 318–324.

66. 'The Riddle of the Sphinx' in Fry (2018), 303.

67. Curry, A. (2013), 'The Milk Revolution', *Nature* **500**, 20–22.

68. Fisher, S. E. and Ridley, M. (2013), 'Culture, Genes, and the Human Revolution', *Science* **340**, 929–930. Ridley has written previously on the interplay of human culture and evolution in his 2003 book *Nature via Nurture: Genes, Experience, and What Makes Us Human*. This book contains the memorable quote 'Professors ascribe the brilliance of their children to Nature and that of their students to Nurture'.

69. A note in *Science* (Curry 2018) comments on the finding that the Mongolians have been dairying for at least 3,000 years, yet 95 per cent of the population is lactose intolerant. The solution to this apparent paradox is that the milk products they consume consist of a variety of fermented products as well as cheese. In an essay, Konrad Bloch (1994) explains why this works: 'Saure Milch was a fairly common item on my family's menu. A dish of milk was left on a windowsill in the summertime. After a few days in the open air, a yellowish semisolid skin, or layer, covered the surface. When sweetened, sour milk made an excellent meal—or so we were told by our parents. Obviously, the milk did not simply spoil, as it does in the refrigerator once the container is opened. The explanation is that airborne bacteria, *Lactobacillus* species, thrive on milk, fermenting the milk sugar lactose to lactic acid, and souring [sterilizing] the liquid. Once these bacteria have taken hold in their favourite medium, they grow vigorously and leave other airborne microbes no chance to compete.' Curry, A. (2018), 'Early Mongolians ate dairy but lacked the gene to digest it', *Science* **362**, 626.

70. A simple experiment helps to illustrate how ATP synthesis in mitochondria depends on formation of a proton gradient. This consists of inserting purified ATP synthase together with ADP, Pi, and bacterial *rhodopsin*, which is a light-driven proton pump, into an artificial membranous vesicle or *liposome*. (See entry for **LIPOSOMES**.) In the presence of light, rhodopsin pumps protons into the vesicle, creating a proton gradient which results in ATP synthesis as protons flow out of the vesicle through the membrane-bound ATP synthase.

71. Perutz (2003), 151.

72. Stevenson, R. D. and Wasserug, R. J. (1993), 'Horsepower from a horse', *Nature* **364**, 195.

73. Steinmetz, P. R. L. (2012), 'Independent evolution of striated muscles in cnidarians and bilaterians', *Nature* **487**, 231–234.

74. Prufer, K. et al. (2014), 'The complete genome sequence of a Neanderthal from the Altai Mountains', *Nature* **505**, 43–49.

75. For details of the procedures involved in transplantation of the synthetic chromosome, see Lartigue, C. et al. (2007), 'Genome transplantation in bacteria: Changing one species to another', *Science* **317**, 632–638. An account of the overall project is in Gibson, D. G. et al. (2010), 'Creation of a bacterial cell controlled by a chemically synthesized genome', *Science* **329**, 52–56.

76. Baker, M. (2013), 'The 'OMES Puzzle', *Nature* **494**, 416–419.

77. Pennisi, E. (2013), 'Opsins: Not Just for Eyes', *Science* **339**, 754–755.

78. Johnson, S. C. and McClelland, S. E. (2019), 'Watching cancer cells evolve', *Nature* **570**, 166–167.

79. Kacharoo, A. H. et al. (2015), 'Systematic humanization of yeast genes reveals conserved functions and genetic modularity', *Science* **348**, 921–5.

80. Emmons, S. W. (2015), 'The beginning of connectomics: a commentary on White et al.' (1986) 'The structure of the nervous system of the nematode *Caenorhabditis elegans*', *Phil. Trans. R. Soc. B* **370**, 20140309. The paper by White et al., also known as 'The mind of the worm', runs to 340 pages.

81. Karlson, P. and Luscher, M. (1959), 'Pheromones: a new term for a class of biologically active substances', *Nature* **183**, 55–56.

82. Ridley, M. (2004), *Nature* **431**, 244.

83. The pK_a of a weak acid HA, such as acetic that dissociates into $H^+ + A^-$, is the pH at which it is half dissociated, i.e. when $[A^-] = [HA]$. Titration of a weak acid with sodium hydroxide generates a sigmoidal titration curve, and the midpoint or inflection point is the pK of the acid. Approximately 1 pH unit on either side of the pK value is the region of useful pH buffering capacity.

84. Where the dark reactions of photosynthesis are concerned, there are two types of plants, termed C3 and C4, the difference between them being that the carboxylation reaction carried out by rubisco is much more efficient in

C4 plants. In C3 plants, rubisco carries out two different reactions: one is the carboxylation of ribulose 1,5-bisphosphate to form a hexose that is converted to the C3 compound 3-phosphoglycerate. The second reaction, *oxygenase*, leads to the formation of a C2 compound, *glycolate* $HOOC-CH_2OH$. Further reactions produce CO_2 and NADH, the latter channelling into the respiratory chain. For C3 plants, the oxygenase and subsequent reactions are referred to as *photorespiration*. In C4 plants, CO_2 is concentrated into a novel *bundle sheath cell* in the form of the C4 compound *oxaloacetate*, which is then decarboxylated. The result is that CO_2 is now at nearly ten times the concentration occurring in C3 plants, the rubisco oxygenase is completely suppressed, and the dark reactions of photosynthesis are correspondingly favoured. C4 plants are favoured by hot and dry climates. Examples of C4 plants are sugar cane, sorghum, and millet.

85. The source of colchicine is the crocus *Colchicum*, a name which derives from the Greek **kolchikon** meaning 'from Colchis', an area on the eastern shore of the Black Sea. This was the destination of Jason and the Argonauts in their mythical quest for the Golden Fleece. At Colchis, Jason marries Medea but later abandons her to marry Glauce. To avenge Jason's betrayal, Medea poisons her two children by Jason, and Glauce. The association between Colchis, Medea, and poison was an enduring myth in classical times and was the story told in the play *Medea* by Euripides, first performed in Athens in 431 BCE.

86. Uhlen, M. et al. (2015), 'Tissue-based map of the human proteome' *Science* **347**, 394 (summary). See the full article at http://dx.doi.org/10.1126/science.1260419. This paper reports results of analyses on major tissues and organs (n = 44) using 24,028 antibodies corresponding to 16,975 protein-coding genes. This was complemented by RNA sequencing data for thirty-two tissues.

87. '*OED2* defines chaperone as a "person, usually a married or elderly woman who, for the sake of propriety, accompanies a young unmarried woman in public as guide and protector." Thus, the traditional role of the human chaperone, if described in biochemical terms, is to prevent improper interactions between potentially complementary surfaces.' Ellis, R. J. and van der Vies (1991), 'Molecular Chaperones', *Ann. Rev. Biochem.* **60**, 330.

88. Crosland (1962), 292.

89. Holmes (2009), 449.

90. Sedgwick (1785–1873) was Professor of Geology at Cambridge, and Darwin, who as a student accompanied Sedgwick on geological field trips, confessed to having never attended his lectures. While they remained lifelong friends, Sedgwick predictably did not like *The Origin of Species*—'utterly false' and 'from first to last it is a dish of rank materialism, cleverly cooked and

served up'.

91. Thomas Henry FRS (1734–1816), founder of the Manchester Literary and Philosophical Society, pronounced the pursuit of natural philosophy, i.e. science, to be preferable to 'the tavern, the gaming table, or the brothel', quoted in Porter (2001), 427.

92. Labunskyy, V. M. et al. (2014), 'Selenoproteins: molecular pathways and physiological roles', *Physiol. Rev.* **94**, 739–777. This paper provides a comprehensive review of *selenoproteins* and of the mechanisms involved in Sec incorporation. See figure 9 for a listing of the twenty-five selenoproteins found in humans.

93. Scudellari, M. (2017), 'To stay young, kill zombies', *Nature* **550**, 448–450. Van Deursen, J. M. (2019), 'Senolytic therapies for healthy longevity', *Science* **364**, 636–637.

94. Green, E. D., Rubin, E. M., and Olson, M. V. (2017), 'The future of DNA sequencing', *Nature* **550**, 179–181.

95. The alternate names for the neuropeptides arose because two groups of researchers, unknown to each other, came up with similar results. One group, interested in the role of the peptides in narcolepsy, christened them *hypocretins*, hypo- from *hypothalamus*, -cretin because of a structural similarity to a gut hormone *secretin*; the other group was interested in their possible role in the modulation of appetite and called them *orexins* after **orexis** (Gk.) 'appetite, desire, longing'.

96. Nicholls (2018). This book provides a comprehensive account of narcolepsy by someone with first-hand experience of the condition. See also Liblau, R. S. (2018), 'Put to sleep by immune cells', *Nature* **562**, 46–48. This discusses the finding that a particular HLA gene is present in 98 per cent of people with narcolepsy, compared to only 15–30 per cent of the general population, depending on ethnicity. See **IMMUNITY** and **AUTOIMMUNE DISEASES** for more on HLA genes.

97. Davy quite reasonably suggested the metals be named for the materials from which they were isolated. Contending mediaeval Latin names were **natrium** and **kalium**, which were current in Germany and championed by Berzelius, who claimed to have etymology on his side: *natrium* was thought to be derived from the Greek **nitron**, which was associated with soda, while *kalium* was derived from the Arabic **al-qili**, translated as *alkali*. Controversies over standardization of the symbols used to represent elements continued for years, and here Berzelius had a win with the symbols Na for sodium and K for potassium.

98. Cyranoski, D. (2018), 'The cells that sparked a revolution', *Nature* **555**, 428–430.

99. Kiviet, D. J. et al. (2014), 'Stochasticity of metabolism and growth at the single-cell level', *Nature* **514**, 376–379.

100. The Thatcher quote makes more sense when placed in context. At that time, she was under pressure to reverse policies that had led to high unemployment. Also, it was a reference to a well-known 1948 play by Christopher Fry: *The Lady's Not for Burning.* 'To those waiting with bated breath for that favourite media catchphrase, the "U-turn", I have only one thing to say: You turn [U-turn] if you want to. The lady's not for turning.'

101. Patel, A. et al. (2017), 'ATP as a biological hydrotrope', *Science* **356**, 753–756.

102. Nadege, P. et al. (2013), 'Pandoraviruses: Amoeba viruses with genomes up to 2.5 Mb reaching that of parasitic eukaryotes', *Science* **341**, 281–285. Dance, A. (2021), 'The incredible diversity of viruses', *Nature* **595**, 22–25.

103. Servick, K. (2017), 'Xenotransplant advances may prompt human trials', *Science* **357**, 1338–1339.

104. Wang, E. Y. et al. (2021), 'Diverse functional autoantibodies in patients with COVID-19', *Nature* **595**, 283–288.

105. Watsa, M. et al. (2020), 'Rigorous wildlife disease surveillance', *Science* **369**, 145–147.

BIBLIOGRAPHY

Alexander, C. *The War that Killed Achilles: The True Story of Homer's Iliad and the Trojan War.* Penguin Books. 2009.

Andrews, E. *A History of Scientific English.* Richard R. Smith, New York. 1947

Augustine St. *Confessions.* Penguin Classics, London. 1961.

Ball, P. *Life's Matrix.* University of California Press, Berkeley. 2001

Ball, P. *Bright Earth: The Invention of Colour.* Penguin Books, London. 2006.

Barnes, J. *Something to Declare.* Picador, London. 2002.

Bloch, K. *Blondes in Venetian Paintings, the Nine-Banded Armadillo, and Other Essays in Biochemistry.* Yale University Press, New Haven. 1994.

Brown, R. W. *Composition of Scientific Words.* Smithsonian Books, Washington. 1956.

Bowman A. K. *Life and Letters on the Roman Frontier: Vindolanda and Its People.* British Museum Press, London. 3rd ed. 2003.

Bragg, M. *The Adventure of English: The Biography of a Language.* Hodder & Stoughton. 2003.

Bryson, B. *Mother Tongue.* Penguin Books. 1991.

Burnside, J. *Word Watching.* Scribe, Melbourne. 2009.

Crosland, M. P. *Historical Studies in the Language of Chemistry.* Harvard University Press, Cambridge, Massachusetts 1962.

Davis, D. M. *The Compatibility Gene.* Penguin, London. 2014.

Davis, D. M. *The Beautiful Cure.* The Bodley Head, London. 2018.

Dixon, M. in *The Chemistry of Life: Lectures on the History of Biochemistry* Needham J. (ed.) Cambridge University Press, Cambridge. 1970.

Edgar, J. D. M. *Immunology.* Elsevier, Edinburgh. 2006.

Fry, S. *Heroes. Volume II of Mythos.* Penguin, UK. 2018.

Heather, P. *The Fall of the Roman Empire.* Oxford University Press, Oxford 2006.

Holmes, R. *The Age of Wonder: How the Romantic Generation Discovered the Beauty and Terror of Science.* Harper Press, London. 2009.

Jackson, B. D. *A Glossary of Botanic Terms, with Their Derivation and Accent.* Gerald Duckworth & Co. Ltd., London. First edition May 1900. Fourth edition, January 1928. Reprinted June 1949.

Jacob, F. *The Statue Within: An Autobiography.* Cold Spring Harbor Laboratory Press, Cold Spring Harbor, New York. 1995.

Kandel E. R. *In Search of Memory: The Emergence of a New Science of Mind.* Norton, New York. 2006. p. 102

Leonhardt, J. *Latin: Story of a World Language.* Belknap/Harvard University Press, Cambridge, Massachusetts. 2013.

Lewis, C. T. and Short, C. *A Latin Dictionary.* Clarendon Press, Oxford. First edition 1879. Reprinted 1933. A much handier but much less comprehensive Latin dictionary is put out by Harper Collins Publishers, first edition 1997.

Liddell, H. G. and Scott, R. *Greek-English Lexicon.* Oxford University Press. Reprint based on the seventh edition. First edition 1889.

Logue, M. and Conradi, P. *The King's Speech.* Quercus, London. (2010).

Mattern, S. P. *The Prince of Medicine: Galen in the Roman Empire.* Oxford University Press, New York. 2013.

Montgomery, S. L. *Does Science Need a Global Language?* University of Chicago Press, Chicago. 2013.

Morwood, J. *Oxford Grammar of Classical Greek.* Oxford University Press, Oxford. 2001

Needham, J. (ed.). *The Chemistry of Life: Lectures on the History of Biochemistry.* Cambridge University Press, Cambridge. 1971.

Nicholls, H. *Sleepy Head: Narcolepsy, Neuroscience, and the Search for a Good Night.* Profile Books, London. 2018.

Perutz, M. F. *I Wish I'd Made You Angry Sooner.* Cold Spring Harbor Press. 2002

Pinker, S. *The Language Instinct.* Penguin Books, London 1994.

Porter, R. *Enlightenment: Britain and the Creation of the Modern World.* Penguin Books. 2001.

Rocke, A. J. *Nationalizing Science: Adolphe Wurtz and the Battle for French Chemistry.* MIT Press, Cambridge, Mass. 2001.

Simpson, J. *The Word Detective.* Little, Brown, London. 2016.

Stearn, W. T. *Botanical Latin.* Fourth edition. David and Charles, Devon, UK 1992

Vogel, S. *Prime Mover: A Natural History of Muscle.* Norton, New York 2001.

Wapner, J. *The Philadelphia Chromosome.* The Experiment, LLC, New York. 2013

Watson, J.D. *The Double Helix.* Penguin Books. 1970.

Yarus, M. (2010), *Life from an RNA World.* Harvard University Press, Cambridge, Mass.

INDEX

A

absolute alcohol 140, 241
absolute temperature 208, 241, 269
absorb, absorbance 37, 67-8, 115, 211, 222, 256-7
abstract 53
accurate 56
ACE (angiotensin converting enzyme) 93
acetabulum 68
acetaldehyde 191
acetic acid 68
acetone 101, 229
acetonitrile 121
acetyl CoA 68-9, 78, 113, 124, 176, 245, 255
acetyl CoA carboxylase 69
acetylated hexosamine 184
acetylation 75, 199
achiral 111
achondroplasia 106
acid hydrolases 102, 219
acid hydrolysate 75
acidic amino acids 75
acidify 53
acidosis 68
acral 29
acridine 41, 68
acromegaly 29
acrosome 30
acrylamide 68
ACTH 69
actin 126, 185, 217
action 53
action potential 240-1
activated charcoal 32, 67

activation cascade 165, 226
activation energy 80, 135, 147
active site 135, 139
active transport 67
actomyosin 126
acute 53, 115
acyclics compounds 72
adaptive immunity 91, 154, 161
adenine, adenosine 195
adenyl cyclase 156
adhere 53
adipocytes 69
adipokines 170
adipose tissue 69-70, 170
adrenal cortex 70
adrenal gland 38, 70, 114
adrenalin 38, 69-71, 156
adrenergic receptors 70
adult stem cells 242
aerobic, anaerobic 54, 167
aetiology 54
affinity chromatography 115
agglutinin 53
aggregation 72, 228
aggresomes 228
agonist 53, 156, 229-30
agra 127
AIDS, HIV 29, 139, 250, 253
alanine, pyruvate 74, 77, 110, 156, 176
albumin 89, 91, 119, 141
alchemy 109, 133
alcohol 45, 49, 136, 140, 159, 173
alcoholic fermentation 262
aldehyde 49, 73-4, 101-2, 211
aldohexose 49, 102
aldosterone 70, 132
alexia 29, 132

E

H

M

mucopolysaccharides 184-5
mucosal tissues 159
mucus 184
multi- 44
multinucleate cells 169, 186, 219
multiple myeloma 183
multiple sclerosis 164, 192
multipotent stem cells 44, 242
muriatic acid 215
mus musculus 185, 213
muscle 129, 185
muscle wasting 82, 269
muscular tissue 246
mutagen 189
mutation 106, 179, 189-90
mutualism 245
myalgia 80, 185
Myasthenia gravis 185
mycelium 191
mycobacteria 191
mycoplasma 191, 223, 254
mycorrhiza 192
mycotoxin 191
myelin 192
myelinated axons 192
myeloblasts 90
myeloid precursor cells 90
myeloperoxidase 46
myoblasts 186
myocytes 185
myofibrils 185
myoglobin 185
myopathies 185, 217, 248
myosin 126, 185, 249
myristic acid 141-2
myristoylation 142, 168

N

N-glycosidic bond 189, 196
N-linked glycans 76, 157
N-terminal amino acid 142, 246

NAD^+, $NADP^+$ 124, 138, 222, 255
NADH, NADPH 83, 125, 181, 221-3, 273
name 192
nano- 24, 44
nanoparticles 44, 94
narcolepsy 85, 239
narcosis, narcotics 239
narrow 60
nascent proteins 194, 228
native 194
natively unfolded proteins 194
natural killer (NK) cells 164
Neanderthal genome 179
near 60
necroptosis 32, 57, 95
necrosis 32, 57, 95
nematode 213
neo- 44
neonates 24, 69
neoplasia, neoplasm 24, 44, 223
neoteny 44
nephropathy 217
nerve impulses 168, 211
nervous tissue 36, 38, 173, 246
neuralgia 80
neuraminidase 88, 157
neurodegenerative diseases 239, 251
neurotransmitters 49, 70, 229, 240
neurotropic 122, 250
neutrophils 46, 91
Newton 109, 119, 267
NHEJ system 155
niacin (Vitamin B3) 127, 255
night 61
nitrogen 80, 110
nitrogenase 80, 138
nocebo 65, 194
nociceptor 195
nomenclature 194
nominative determinism 171

O

repair mechanisms 190
repeating disaccharides 184
replication errors 189-90
replicative immortality 98
respiration 54
respiratory chain 273
reticulum 130, 134
retina 211
retinal, retinol 168, 211, 256
retinal innervation 117
retrogene 267
retrotranscription 106, 267
retrotransposons 149-50
retrovirus 139, 253
reverse complement 198
reverse transcriptase 149-50, 198-9,
    253, 262, 267
reverse turn 227
rheumatism 144
rhinitis 61
Rhizobia 80, 245
rhodopsin 211
riboflavin (vit. B2) 255
ribonuclease 227
ribonuclease P 268
ribonucleotide reductase 83, 167
ribose 103, 195, 261
ribosomal RNA 129, 136, 181, 198,
    200, 268
ribosome 77, 136, 157, 200, 228
ribozymes 135-6
right-handed DNA helix 196
RNA 136, 195, 200, 221, 252
RNA-dependent RNA polymerase
    198, 252
RNA-DNA hybrid 39, 155
RNA virus genomes 253
RNA viruses 198, 252
RNAPs, Pol I, Pol II, Pol III 198-9
rod cells 168, 211
role of sleep 240

rotavirus 166, 250
rough ER, smooth ER 134
rubisco 124, 223, 273
rumen 130

S

S-adenosylmethionine 77
S phase 181
Saccharomyces cerevisiae 152, 191, 213,
    262
Sahul 152-3
saline, salt 232
salivary amylase 179-80
salt linkage 75, 197
saponification 171
sarcocentric 187
sarcoma 186
sarcophagus 186, 188
sarcoplasm 185
sarcoplasmic reticulum 134, 185
sarcosine 186
SARS Covid-2 54, 93, 263, 275
schedule 62
Scheele 214
Schiff base 211, 255
science, scientist 8, 61, 171-2, 233-4
scientific knowledge 232, 234
scope 44, 66, 234
scurvy 85, 255
sebaceous glands, sebum 113
Sec insertion sequence 237
second messenger 124, 156, 202, 211
secondary symbiosis 83
secreted proteins 134, 226
secretin 116, 274
secretion mechanisms 235
secretome 226
sedimentation coefficient 107, 229
segmented genome 252
selenium 78, 236, 245
selenocysteine 78, 237